Renewable Energy
& Sustainable Design

Scott Grinnell

CENGAGE
Learning·

Australia • Brazil • Japan • Korea • Mexico • Singapore • Spain • United Kingdom • United States

CENGAGE Learning™

Renewable Energy & Sustainable Design
Scott Grinnell

VP, General Manager, Skills and Planning: **Dawn Gerrain**

Director, Development, Career and Computing: **Marah Bellegarde**

Senior Product Development Manager: **Larry Main**

Associate Product Manager: **Nicole Sgueglia**

Senior Content Developer: **Jennifer Starr**

Product Assistant: **Scott Royael**

Marketing Manager: **Scott Chrysler**

Senior Production Director: **Wendy Troeger**

Production Manager: **Andrew Crouth**

Content Project Management: **S4Carlisle Publishing Services**

Art Direction: **S4Carlisle Publishing Services**

Media Editor: **Deborah Bordeaux**

Cover image(s): **© Elenathewise; iStock**

Library of Congress Control Number: 2013957888

ISBN: 978-1-1115-4270-2

Cengage Learning
20 Channel Center Street
Boston, MA 02210
USA

Cengage Learning is a leading provider of customized learning solutions with office locations around the globe, including Singapore, the United Kingdom, Australia, Mexico, Brazil, and Japan. Locate your local office at: **www.cengage.com/global**

Cengage Learning products are represented in Canada by Nelson Education, Ltd.

To learn more about Cengage Learning, visit **www.cengage.com**

Purchase any of our products at your local college store or at our preferred online store **www.cengagebrain.com**

Notice to the Reader

Publisher does not warrant or guarantee any of the products described herein or perform any independent analysis in connection with any of the product information contained herein. Publisher does not assume, and expressly disclaims, any obligation to obtain and include information other than that provided to it by the manufacturer. The reader is expressly warned to consider and adopt all safety precautions that might be indicated by the activities described herein and to avoid all potential hazards. By following the instructions contained herein, the reader willingly assumes all risks in connection with such instructions. The publisher makes no representations or warranties of any kind, including but not limited to, the warranties of fitness for particular purpose or merchantability, nor are any such representations implied with respect to the material set forth herein, and the publisher takes no responsibility with respect to such material. The publisher shall not be liable for any special, consequential, or exemplary damages resulting, in whole or part, from the readers' use of, or reliance upon, this material.

Printed in United States of America
Print Number: 01   Print Year: 2014

Table of Contents

CHAPTER 8   Wind Power   193

CHAPTER 9   Hydropower   219

Preface

Renewable Energy & Sustainable Design is an introductory textbook that offers a fundamental understanding of common forms of renewable energy and the principles and methods of sustainable building design. The text requires no science or mathematics prerequisites other than basic algebra and can be covered in a semester-long general education course at either the high school or the college level.

The book begins with an overview of green building practices, addressing principles, materials, and designs that together can create comfortable structures and allow individuals and communities to live healthfully and productively. The book continues by introducing energy, the physical laws that govern its use, conventional power generation, and common forms of renewable energy. The last topic occupies six chapters, each of which covers a different form of renewable energy. Beginning with a brief historical background, these chapters examine the basic science and current technology utilized to generate energy and then continue with an evaluation of the benefits, limitations, and environmental impacts.

The integration of sustainable building design with renewable energy is intended to interconnect our daily lifestyles and the homes we live in to broader concerns over climate change and an uncertain energy future. The subject matter is pertinent to a broad audience, and the nontechnical approach makes the text accessible to individuals of diverse backgrounds and interests. Applicable for coursework at high schools, community colleges, and four-year institutions, the text can be used for majors in environmental science, sustainable community development, physical science, and nonscience disciplines throughout a liberal education curriculum. The text's easy-to-read, factual approach encourages students to reconsider preconceptions and make better-informed decisions about these topics.

Renewable energy and building science are multifaceted topics, and both have experienced dramatic changes over the past decade. This book offers an overview of the benefits and drawbacks of renewable energy systems and building designs using general scientific principles, while attempting to avoid overly complex or technical material. Each chapter concludes with a set of review questions that emphasize the basic principles and important ideas, followed by problems that apply the concepts mathematically.

Objectives of This Book

Human activities, long considered to be insignificant when compared to the vastness of nature, are now recognized as being able to change the world in profound ways. Apprehension over climate change, limited energy supplies, air and water pollution, and a broad array of environmental and human health concerns is now widespread. This book attempts to address the fundamental aspects of human shelter and energy supply, and to serve as a platform for investigating the greater subject of creating a globally sustainable society. The book provides information on a personal scale whenever possible, using real-life case studies and examples so that students may better translate the information into actions they can take. Ultimately, this book is intended to encourage a hopeful perspective for the future that empowers solutions to our global concerns.

Almost daily advances in technology offer abundant opportunities for incorporating current events to complement the foundations established in this book. Emerging technology—ranging from innovative building materials, to improvements in solar hot water systems, to new and potentially revolutionary photovoltaic devices, to ruffled wind turbine blades patterned after whale fins, to advances in gasification processes, to cellulosic ethanol from switchgrass—presents valuable opportunities to supplement the text and foster deeper discussions. This book, therefore, is intended to serve as the starting point for many different conversations, to raise a variety of questions, and to present ideas that can be further investigated according to interest.

Organization of This Book

Chapter 1 introduces the six principles of green building that serve as guidelines for creating efficient, healthful structures that improve occupant comfort, enhance communities, and safeguard the environment. Chapter 2 goes on to investigate common building materials, comparing the energy and resources required for their manufacture, their levels of toxicity, and other factors in order to determine the benefits, limitations, and overall suitability of these materials for various building applications. Chapter 2 also conveys principles of heat transfer and uses these principles to compare types of insulation and assess energy efficiency. The *Further Learning* box in Chapter 2 applies an equation for thermal conduction using R-values to assess heat transfer in buildings.

Chapter 3 focuses on creating comfortable and energy-efficient structures through passive solar design—the ability to heat and/or cool buildings using energy available naturally from the predictable motions of the sun. The discussion of building designs concludes with alternative construction (Chapter 4), a mix of traditional and recent techniques developed to create spaces that feel more natural.

Chapter 5 provides a basic understanding of energy. It begins by establishing solar energy as the foundation of nearly all energy on earth, then provides an overview of fossil fuels and renewable energy, and continues with essential thermodynamic laws. Finishing up with a discussion of solar resource, Chapter 5 sets the stage for solar hot water systems (Chapter 6) and photovoltaic systems (Chapter 7).

Rapid growth and interest in wind power (Chapter 8) position it immediately after solar energy, followed by a discussion of hydropower (Chapter 9). Chapter 10 investigates the assorted methods of capturing usable energy from biomass—a broad category of materials ranging from agricultural and forestry residue, to animal manure, to oilseed crops, to municipal solid waste and landfill gas. Chapter 11 concludes the book with an overview of the two forms of renewable energy not derived from solar energy: tidal and geothermal. The emerging technology of creating low-temperature geothermal power by hydraulic fracturing, derived from the oil and gas industry, exemplifies one of the many starting points for deeper investigation.

Features of This Book

Each of the 11 chapters contains one or more features intended to enhance student learning. These features expand on material presented in the main text, explore a closely related subject, or encourage students to assimilate ideas from previous chapters. These features include:

- *A Full-Color and Visually Intensive Design* that serves to focus student attention and bring relevance to textual descriptions.
- *Further Learning Boxes* that expand on information presented in the main text and offer increased depth for those students who wish to explore the topic in greater detail. These include numerical calculations of heat transfer (Chapter 2), passive annual heat storage (Chapter 4), ocean current energy technology (Chapter 8), early waterwheels (Chapter 9), ethanol in transportation (Chapter 10), and the physics of two daily tides (Chapter 11).
- *Examples* that provide students with numerical calculations using the relationships and equations presented in the text. These examples assist students in finding solutions to the problems given at the end of each chapter and demonstrate helpful problem-solving techniques.
- *Case Studies* that investigate real-life situations and exemplify the ideas presented in the text. These include stick huts built by the Kuna Indians (Chapter 1), common methods of construction (Chapter 2), passive solar design (Chapter 3), a rammed-earth tire structure (Chapter 4), a solar hot water system (Chapter 6), photovoltaic arrays (Chapter 7), commercial-scale wind power (Chapter 8), and micro hydro systems (Chapter 9).
- *End-of-Chapter Review Questions* that encourage students to synthesize concepts and assimilate material from previous chapters. Some of the questions foster discussions, many are open-ended, and others encourage students to evaluate preconceptions or personal biases. Many of these questions can be used to engender lively classroom debates.
- *End-of-Chapter Problems* that test a quantitative understanding of the material and require computation. Useful for homework assignments, group projects, or quizzes, these vary in difficulty but require nothing more than basic algebraic skills.
- *A Comprehensive Glossary* that summarizes key terms and concepts. Important terms are bold-faced and key ideas italicized throughout the book.

Although the book focuses primarily on applications within the United States (and uses familiar units of feet, gallons, Btus, and degrees Fahrenheit), it acknowledges the interdependence of our global community through the use of international case studies and examples.

Supplements to This Book

An accompanying *Online Companion Site* includes materials for both educators and students to help facilitate and enhance the teaching and learning experience.

- *Lesson Plans* outline the key points in each of the chapters and correlate to the accompanying PowerPoint® presentations for seamless classroom integration.

- *Answers to Questions* provide answers and solutions to the review questions and problems included in the chapters.
- *PowerPoint® Presentations* offer a visual representation of the content, including select full-color graphics and photos from the book to emphasize important points.
- *Cognero,* a flexible online system, offers chapter by chapter test questions, and enables instructors to:
  - author, edit, and manage test bank content from multiple resources;
  - create multiple test versions in an instant; and
  - deliver tests from instructor/institution-specific learning management systems or classrooms.
- *Image Gallery* includes all full-color graphics and photos from the book so that educators may add to the existing PowerPoint® presentations or create their own classroom presentations.

For these instructor-specific resources, please visit CengageBrain.com at http://login.cengage.com, and follow the prompts for obtaining access to this secure site.

Acknowledgments

I am grateful to the many contributors of ideas and photographs used throughout the textbook. The dedication and enthusiasm of these individuals, many of whom have been pursuing satisfying and sustainable lifestyles for years, brought great satisfaction to the project. Most particularly, I would like to thank the people who provided information and imagery used for the various case studies:

John Cannon
Chris Duke
Autumn Kelley
Jeff Meerdink
Michael and Judith Peyton
Lowell Rasmussen
Dave Taggett
Alta Turner
Jerry and Diana Unruh
Brent and Amy Wiersma

At Cengage Learning, I would like to thank Jennifer Starr, Content Developer, and Nicole Sgueglia, Product Manager, who brought the book to publication.

Scott Grinnell

Unit Conversion Factors

LENGTH

1 inch (in) = 2.5400 centimeters (cm)

1 foot (ft) = 0.3048 meter (m)

1 mile = 1.609 kilometers (km)

1 cm = 0.3937 in

1 m = 3.2808 ft

1 km = 0.6214 mile

AREA

$1 \ in^2 = 6.4516 \ cm^2$

$1 \ ft^2 = 0.0929 \ m^2$

$1 \ cm^2 = 0.1550 \ in^2$

$1 \ m^2 = 10.764 \ ft^2$

VOLUME

$1 \ in^3 = 16.387 \ cm^3$

$1 \ ft^3 = 0.02832 \ m^3$

1 gallon (gal) = 3.7854 liters (L)

$1 \ gal = 0.13368 \ ft^3$

$1 \ cm^3 = 0.061024 \ in^3$

$1 \ m^3 = 35.315 \ ft^3$

1 L = 0.26417 gal

WEIGHT

1 pound (lb) = 0.45359 kilogram (kg)

1 gal water = 8.34 lbs (at 50°F)

1 kg = 2.2046 lbs

1 gal water = 3.78 kg (at 10°C)

SPEED

1 mile/hr (mph) = 0.44705 meter/second (m/s)

1 mph = 1.6092 kilometers/hour (km/hr)

1 ft/s = 0.68182 mph

1 m/s = 2.2369 mph

1 km/hr = 0.62137 mph

1 m/s = 3.6000 km/hr

DENSITY

$1 \ lb/ft^3 = 16.0185 \ kg/m^3$

$fresh \ water = 62.4 \ lbs/ft^3$

$sea \ water = 78.0 \ lbs/ft^3$

$1 \ kg/m^3 = 0.06243 \ lb/ft^3$

$fresh \ water = 1000 \ kg/m^3$

$sea \ water = 1025 \ kg/m^3$

ENERGY

1 calorie = 4.186 joules (J)

1 British thermal unit (Btu) = 1055.1 J

1 Btu = 0.00029307 kilowatt-hour (kWh)

1 J = 0.2389 calorie

1 J = 0.0009478 Btu

1 kWh = 3412.2 Btu

1 kWh = 3,600,000 J

POWER

1 Btu/hr = 0.29307 watt (W)

1 horsepower (hp) = 2544.4 Btu/hr

1 W = 3.412.2 Btu/hr

1 hp = 745.7 W

1 W = 1 J/s

Introduction to Green Building

Courtesy of Scott Grinnell.

Introduction

Buildings have been an essential aspect of human existence since the earliest civilizations, serving as homes, temples, and centers of commerce. Early people began by modifying natural shelters (Figure 1.1). Later, people constructed simple structures from locally available materials (Figure 1.2 and Figure 1.3). As resources became more widely distributed and manufactured products became commonplace, buildings became larger and more complex (Figure 1.4). As a result, buildings have had an increasingly greater impact on the environment, requiring more resources and energy to construct and operate. Many of our current building practices were developed when fossil fuels were inexpensive and abundant, a condition that will not persist into the future.

Today, in the United States, there are more than 76 million residential buildings and 5 million commercial buildings. Each year, on average, the United States constructs more than 1 million new homes.[1] The manner in which these homes are built and operated profoundly influences our human health and the health of our environment (Figure 1.5). Buildings require energy and resources to construct and operate, resources that must be harvested or mined. The extraction of these resources impacts the environment and our quality of life.

Green building is an effort to build responsibly and sustainably. As human population increases and available natural resources dwindle, green building practices become increasingly important for the health of individuals, communities, and the global environment.

FIGURE 1.1 More than 800 years ago, the Anasazi Indians built elaborate dwellings in south-facing cliffs. Constructed of locally gathered stone, wood, and adobe, these shelters—including this one in Ute Mountain Tribal Park—welcomed the warmth of the winter sun but blocked the higher summer sun. *Source:* David Parsons/NREL.

FIGURE 1.2 Plains Indians built tepees for shelters using native fir poles and buffalo skins. Plains Indians migrated frequently and could dismantle and pack their tepees in less than an hour. *Source:* Library of Congress, Prints & Photographs Division, LC-USZ62-104919.

FIGURE 1.3 Traditional log cabins used locally harvested materials that minimized environmental impacts. *Source:* Library of Congress, Prints & Photographs Division, photograph by Carol M. Highsmith, LC-USZ62-104919.

FIGURE 1.4 The skyscrapers of New York City rely on manufactured products like steel, glass, plastic, and concrete that are often mined, processed, and transported from distant regions. The construction of modern buildings consumes vastly more energy and resources than traditional buildings. *Source:* U.S. Department of Housing and Urban Development.

FIGURE 1.5 Housing developments, like this one in the Pacific Northwest, often proceed without careful consideration of design, material use, or environmental impacts. These failures can compromise the health and comfort of the occupants, threaten valuable ecosystems, and degrade whole communities. Courtesy of John Marzluff.

Principles of Green Building

Green building minimizes resource consumption and harmful impacts on the environment and creates healthy living spaces for people. Creating a green-built structure requires a holistic approach, where green elements are deeply embedded in the design and are established long before construction begins. This is accomplished through appropriate siting, land use, design, material selection, construction techniques, operation, and maintenance. The characteristics that make a building green depend on the local climate, existing infrastructure, available resources, and so on. Buildings must be considered in concert with their ecologies and cultures, not as separate entities. A building that could be considered green in one location may not be green in another.

Green building does not impose a particular design or construction method, nor does it generally appear distinctive from standard construction. Green building need not be technologically advanced or expensive to build. Green building simply minimizes resources and environmental impacts, while creating comfortable spaces for its inhabitants. All green-built structures follow six guiding principles. A green building must be:

- comfortable for its occupants,
- healthful for its occupants,
- energy efficient,
- resource efficient,
- long-lived, and
- made to minimize harmful environmental impacts.

Occupant Comfort

A green building must be comfortable for its occupants both physically and psychologi-cally. Factors that determine occupant comfort include temperature, humidity, lighting, acoustics, and a sense of place.

- **Constant temperature:** The range of temperature that is comfortable for most people varies by only 6°F–8°F. For people wearing typical seasonal clothing dur-ing sedentary activities, this comfort range usually spans 69°F–75°F in the winter and 74°F–80°F in the summer, depending somewhat on humidity, as shown in Figure 1.6. A green building is one that maintains a comfortable and stable daily temperature for its occupants.
- **Constant humidity:** Although humidity varies with the season, extreme levels—either too high or too low—adversely impact occupant comfort. Too little humidity can lead to chapped skin and lips, scratchy nose and throat, and difficulty breathing and can per-mit a buildup of static electricity. Too much humidity can lead to musty air and allergic reactions associated with mold and mildew growth. For a temperature of about 68°F, most people find a relative humidity between 30% and 70% to be acceptable.
- **Lighting:** Lighting influences ambience and can greatly affect moods and atten-tion spans of building occupants. Numerous studies have shown a marked increase in worker productivity when occupying buildings illuminated by natural light and

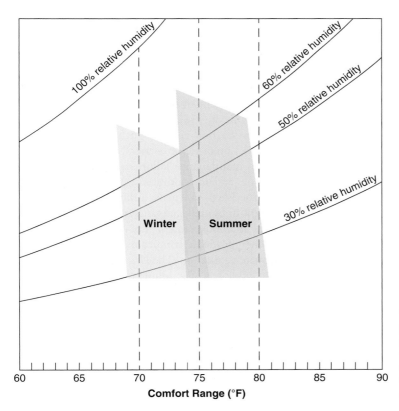

FIGURE 1.6 The shaded regions represent common comfort ranges of temperature and humidity for people clothed in typical winter and summer dress during sedentary activity.

Source: Adapted from Energy Solutions Professionals.

offering views to the outside.[2] Teachers and students alike perform better in classrooms with natural lighting.

- **Acoustics:** The acoustics of a building have a major impact on the productivity, attention span, and stress level of its occupants. Buildings must be designed to permit individuals to carry on conversations comfortably while other sounds are present, as well as provide quiet locations as desired.
- **Sense of place:** Public buildings and workplaces should foster social interaction and a sense of connection with the community. Residential buildings need to satisfy the subjective needs of the occupants, such as feeling safe. Buildings must also permit people to feel connected to natural spaces. Providing windows on two sides of a room enhances a feeling of connection to the outdoors, and windows that open allow occupants to better control their surroundings.

Occupant Health

People living in industrialized countries typically spend 80%–90% of their time indoors. This can have major consequences for occupant health. The U.S. Environmental Protection Agency estimates that more than 30% of all buildings have indoor air quality that is worse than the outside air.[3] Constant exposure to polluted air in buildings is a cause of many health disorders, a condition sometimes called sick building syndrome.

Many common building materials release chemicals that are toxic to people and the environment. Volatile organic compounds (VOCs), such as formaldehyde, benzene, toluene, and perchloroethylene, are present in many common building materials including particleboard, plywood, paints, adhesives, carpets, and plastics. Many of these compounds are known carcinogens that can outgas into buildings, compromising air quality and causing headaches, asthma, depression, dizziness, reduced productivity, immune system disorders, and many other health problems.

High levels of humidity or improper management of water can lead to respiratory and immune system disorders associated with the growth of mold and mildew. Water management problems can arise from construction and design failures such as poor drainage around buildings, improper installation of vapor barriers, and inadequate ventilation. Health issues can also arise from maintenance failures such as not regularly cleaning filters in HVAC (heating, ventilation, and air conditioning) equipment.

Buildings can also become unhealthful due to allergens, airborne bacteria, smoke, pet dander, and dust mites that are brought in by the occupants of the buildings themselves. Odors and fumes from appliances and cook ranges, vapors from cleaning supplies, and VOCs from furniture and other products compromise indoor air quality. Consequently, all buildings require adequate ventilation. The U.S. Department of Labor recommends a ventilation rate of 15–20 cubic feet per minute of fresh air per person within public buildings. Residential buildings should fully exchange indoor air with fresh air at least eight times per day and sometimes considerably more often.[4]

A green building is one that is free of toxins and other harmful agents, provides adequate ventilation, and promotes good health.

Energy

The energy associated with any structure takes on two forms: the initial energy required to construct the building and the ongoing energy necessary to operate and maintain it.

The construction and operation of modern buildings consume a great deal of energy, accounting for a third of all energy used in the United States and two-thirds of all electricity (see Figure 1.7). Most of the energy used in the United States comes from burning fossil fuels, particularly coal (Figure 1.8) and oil (Figure 1.9). The generation of electricity and refinement of oil cause air and water pollution, acid rain, and climate change and have other adverse effects on human health and the environment.

Building Construction

The energy required to construct a building begins with the harvesting and mining of raw materials. Additional energy is used to transport, process, and manufacture the building materials and then to transport the finished products to the building site. Finally, the installation of the materials and construction of the structure require energy. The sum of all of this energy, from harvest to manufacture to installation, is known as the embodied energy of the material. A green-built structure minimizes the total embodied energy of the structure.

Building Operation and Maintenance

In addition to minimizing the energy consumed during construction, a green building is designed and operated so as to use as little energy as possible when occupied. Energy efficiency includes appropriate selections of

FIGURE 1.7 Lights outline the city of Chicago and contrast starkly with the empty water of Lake Michigan. This aerial image reveals the enormous impact of the building industry on the environment. The energy operating our homes and cities comes predominantly from fossil fuels. Image courtesy of the Image Science & Analysis Laboratory, NASA Johnson Space Center.

FIGURE 1.8 An aerial view of a coal mine in Wyoming reveals the environmental impact of open pit coal mining. Coal serves as the fuel source for most of the electricity produced in the United States. © JVvrublevskaya/Shutterstock.com.

FIGURE 1.9 The extraction, refining, transportation, and consumption of oil have large environmental impacts. Some of these impacts include air and water pollution, loss of wildlife habitat, and climate change. © huyangshu/Shutterstock.com.

appliances, lighting, fixtures, and HVAC systems, as well as adequate occupant education on the proper use of these systems. HVAC systems that are too large or too small for a structure waste energy or may fail to keep the building comfortable. A green building also seeks to minimize the use of fossil fuels, substituting renewable energy sources whenever feasible, such as generating electricity with the solar array shown in Figure 1.10.

Heating and Cooling In climates with large seasonal variation, which includes most of the United States, heating and cooling comprise a building's largest energy demands. To maintain a comfortable temperature, while minimizing energy consumption, a green building must be well insulated. Additionally, green-built structures seek to employ low-energy solutions to heating and cooling. For example, designing a building to take advantage of natural solar heating can reduce energy use by 30% in the northern United States. Similarly, natural methods of cooling, such as evaporative cooling towers in desert climates (shown in Figure 1.11) and ceiling fans and open floor plans in more humid climates, can keep a building comfortable without the need for conventional energy-intensive air conditioning units.

Lighting An iconic example of energy efficiency is the replacement of incandescent light bulbs with **CFL** (compact fluorescent) or **LED** (light-emitting diode) bulbs. These bulbs use only a quarter as much energy as incandescent bulbs of the same brightness, and some use even less energy. In buildings where lighting is required most of the time, this switch not only saves energy in operating the building but also reduces the building's total embodied energy and maintenance costs, as CFL and LED bulbs last much longer than incandescent bulbs. Although CFL bulbs contain trace amounts of mercury and must be disposed of properly, their use reduces mercury pollution by consuming less electricity, which is typically generated at mercury-emitting coal-fired power plants.

FIGURE 1.10 The use of renewable energy, such as these solar modules, reduces dependency on fossil fuels. *Source:* Beamie Young/NIST.

FIGURE 1.11 The visitor center at Zion National Park uses evaporation to keep the building cool. As water evaporates in the tower, it cools the air and creates natural downdrafts that flow into the building. *Source:* Robb Williamson/NREL.

Phantom Loads Many common electronic devices, such as televisions, phones, and computers, never fully turn off. Even when switched off, these devices continue to draw electricity and over time can consume a significant amount of energy. To prevent this hidden use of energy, known as phantom loads, these devices must be disconnected from the power supply. One method of accomplishing this is to plug electronic devices into power strips that can be turned off when not in use.

Occupant Education Even buildings that incorporate energy-efficient systems can fail to perform optimally if the occupants do not use the systems correctly or neglect to adopt energy-saving practices. For example, systems operated outside of their designed limits, used when inappropriate, not turned off when no longer needed, or used hazardously will erode the overall energy efficiency of the building.

Resources

The impact of the building industry on earth's limited natural resources is staggering. According to the U.S. Department of Energy, buildings account for approximately 30% of the wood and raw materials consumed, 25% of the water used, and between 20% and 40% of the solid waste generated annually.[5] The material resources for constructing a building must either be harvested, such as by clear-cutting (Figure 1.12), or be mined (Figure 1.13) and then processed into their finished form (Figure 1.14). Each stage consumes energy that up until the present has generally been supplied by fossil fuels, an increasingly limited resource. In addition, large-scale harvesting and mining generally require the construction of roads, disturb wildlife habitat, cause ecological damage, and result in other negative environmental impacts.

FIGURE 1.12 The demands of the building industry drive the clear-cutting of timber, shown here on Vancouver Island, British Columbia.

© iStockphoto.com/Ian Chris Graham.

Green building optimizes the use of materials and minimizes the depletion of natural resources. Efficient designs, careful planning, reuse of construction materials, and extensive recycling can dramatically reduce construction waste. Furthermore, designing a structure to utilize sustainably harvested timber, salvaged and recycled materials, and locally sourced products (to reduce the embodied energy of transportation) further diminishes the depletion of natural resources.

In addition to raw materials, water is another valuable resource. Green building reduces the consumption of water through efficient design, appropriate plumbing, and the use of low-flow fixtures. Other methods of minimizing water use include selecting native landscaping, incorporating systems that collect and store rainwater, and plumbing toilets to filter and reuse sources of grey water (such as sinks and showers). These techniques can substantially cut water consumption over the life of the building.

Another resource-conserving aspect of green building is the use of the smallest, most efficient design that will serve the needs of its occupants. A small structure reduces overall material use and the amount of space that requires heating, cooling, and maintenance. Since 1970, the average size of

FIGURE 1.13 The Bingham Copper Mine, located near Salt Lake City, Utah, has removed more than 17 million tons of ore since its beginnings in 1903. The environmental impact of open pit mining can be extensive. *Source:* Craig Fong/NREL.

FIGURE 1.14 Once ore is mined, crushed, and prepared for smelting, it is transported to mills, where energy-intensive processes convert the refined ore into marketable metal. © dominique landau/Shutterstock.com.

residential homes has steadily increased, resulting in homes that are often larger than necessary or practical for their occupants. According to the U.S. Census Bureau, in 1970, the average home size was 1400 square feet; by 2007, that value had peaked at 2521 square feet. Recently, this trend has leveled off and even reversed somewhat, a positive sign for reducing resource consumption.[6]

Longevity

Green buildings are long-lived compared to the energy and resources of construction. The lifespans of different styles of shelters can vary dramatically, from grass huts and buffalo skin teepees to homes built of stone and concrete. While a house built of stone and concrete may outlast a grass hut, both might satisfy the longevity principle because of the enormous difference in resource consumption between the two styles. Conversely, a poorly designed suburban home that suffers irreparable damage during a modest windstorm would not satisfy the longevity principle.

In addition to being long-lived, green buildings should be able to be recycled at the end of their useful life. Materials salvaged from old buildings provide valuable resources for new construction and reduce the demand on raw materials.

Environmental Impacts

Buildings impact the environment in a variety of ways other than the consumption of energy and resources. These include fragmenting natural habitats, disrupting hydrological systems, and impacting ecosystems through pollution and changes to nutrient cycles. Table 1.1 indicates common building impacts and offers possible solutions.

TABLE 1.1 Environmental impacts.

ISSUE	EXAMPLE SOLUTIONS
Storm water runoff that can lead to erosion and sedimentation of nearby streams and lakes	Permeable paving, living roofs and walls, appropriate landscaping, and vegetative barriers significantly decrease problems arising from storm water runoff.
Chemical pollution from fertilizers, pesticides, preservatives, paints, and cleaning agents	Landscaping with native plants that are well suited to the local climate limits the application of chemical fertilizers and pesticides, reduces labor, and lowers maintenance costs. Select building materials that do not require harmful chemicals for cleaning and maintenance.
Nighttime light pollution from exterior (and some interior) light fixtures	Shielded fixtures that direct light down where it is needed not only prevent light pollution of the night sky but also reduce energy costs by more efficiently illuminating the intended areas.
Heat-absorbing materials that give rise to a "heat island effect," where the buildings become warmer than their surroundings	Install light-colored or living roofs rather than dark asphalt or tar roofs. Avoid large paved areas that heat rapidly in the sunlight.
Degradation of wetlands and other sensitive habitats	Avoid building in areas that adversely affect sensitive ecosystems.

Building Siting

The location of a building is as important as its design. Buildings have the least impact on the environment when they are located close to existing services and infrastructure, such as shopping centers, grocery stores, medical care facilities, entertainment, schools, and places of employment. If a commute is necessary, then building close to bus routes and commuter lines reduces car dependency and the energy associated with day-to-day travel.

Where possible, buildings should reuse sites that have been previously built on rather than developing undisturbed land. Construction should be avoided on wetlands, flood plains, or unstable soil; on usable agricultural land; in regions with sensitive ecosystems; or in areas of archaeological, historical, or cultural significance. Figure 1.15 shows an example of a housing development that compromises the original ecosystem and encourages excessive consumption of energy and resources through commuting.

The building site should take into account the direction and proximity of factories, airports, highways, and other centers of noise and pollution. The site should enable a building to be optimally oriented with respect to the path of the sun to maximize passive heating (in cold climates) or cooling (in warm climates). Finally, the site should allow the building to harness available renewable energy, such as solar and wind.

FIGURE 1.15 Residential housing complexes, like this one built on wetlands near Miami, Florida, often require residents to commute by private automobiles. Green building practices advocate the selection of building sites away from wetlands and other sensitive ecological areas.

© istockphoto.com/Jodi Jacobson.

Climate Considerations

Historically, people created shelters applicable to their specific location, not only using local materials but also selecting designs suitable to their climate. Technical advancements during the Industrial Revolution resulted in standardized building materials and encouraged similar designs and construction techniques across differing climates. This led to the construction of buildings that were inefficient for their location and consumed more energy and resources than climate-appropriate structures. Designing a home suitable to the local climate is essential to green building. Consider traditional residential designs characteristic of four different climates (also shown in Figure 1.16).

- **Cold climate design** (e.g., sub-Arctic Canada, northern Scandinavia): Minimizing exposure and heat loss is essential in cold climates. Successful strategies include using compact designs with minimal surface area, partially submerging the structure underground where the mass of earth moderates temperature extremes, making use of living roofs to supply additional insulation, and landscaping to protect against the winter wind. Windows are few and located only on the sunward side (south for the Northern Hemisphere), and walls, ceilings, and floors are heavily insulated.

- **Temperate climate design** (e.g., northern United States, southern Canada): In a climate that is cold in the winter and warm in the summer, buildings should be well insulated and oriented with the long side facing sunward to allow better solar heating in the winter. Window size and location should be well considered, with most on the sunward side. Although windows on other sides of the building lose more heat than they provide during cold weather, they may be necessary to provide adequate daylight and a sense of connection to the outdoors. Landscaping should protect from the winter wind and offer shade from the summer sun.

- **Hot, arid climate design** (e.g., southwestern United States): A successful building strategy for regions that become hot during the day and cold at night includes compact designs with thick, massive walls that moderate daily temperature swings. This climate encourages selective placement of windows and eaves and the use of landscaping that prevents direct illumination of living spaces by the sun during the summer months. North-opening U-shaped buildings with open, shaded courtyards have long been an effective and popular design in the American Southwest. Fountains and garden pools are beneficial if water resources permit. Designs that provide ample shade for the occupants and employ natural evaporative cooling instead of electrically operated air conditioning can significantly reduce energy consumption.

- **Hot, humid climate design** (e.g., Florida and tropical locations): Buildings constructed in hot, humid climates perform better when raised to catch the breeze and designed with an open floor plan that allows plenty of air movement. Landscaping needs to shade the building but still permit air motion. Wind traps and vents can funnel breezes into living spaces, supplemented by energy-efficient ceiling fans. When relying on natural cooling, these buildings generally do not require significant insulation because the daily and annual temperatures vary only slightly. For the same reason, lightweight building materials perform as well as more massive

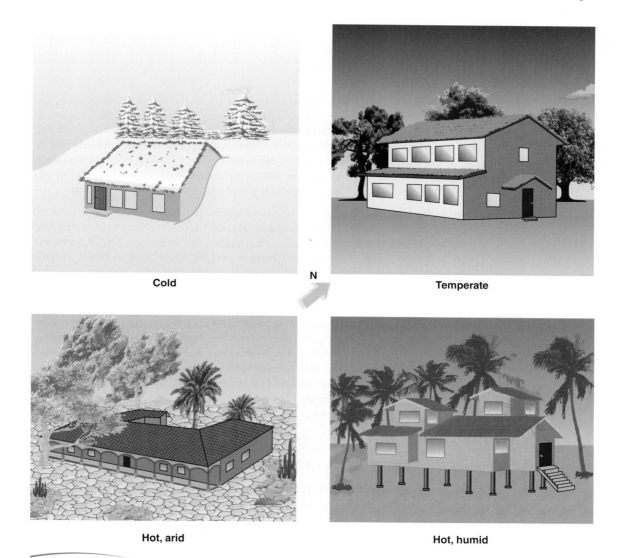

Cold

N

Temperate

Hot, arid

Hot, humid

FIGURE 1.16 Different climates drive different building priorities and require different designs and landscaping to optimize performance. © 2016 Cengage Learning®.

materials. The orientation of the building is often a compromise between optimal shading and best air flow, though windows on the east and west sides should be minimized to prevent excess solar heating unless they can be well shaded. Isolating cooking spaces from the sleeping spaces and incorporating outside living spaces can also make the building more comfortable.

CASE STUDY 1.1

Kuna Indians

The Kuna people live on the San Blas islands off the coast of Panama. Surrounded by the Caribbean Sea, these coconut-crowned islands experience a comfortable tropical climate all year. The Kuna tend fields on the mainland where they grow crops and obtain firewood and drinking water. Materials harvested from the mainland make up their homes: unmilled poles for framing, sticks for walls, and palm thatching for roofs. The Kuna cook on open pit fires, sleep in hammocks, and walk across sand floors, which they share with curious crabs and other sea creatures. These traditional structures uphold many of the principles of green building.

- **Comfortable:** Without insulation or air conditioning, the temperature and humidity within Kuna homes remain relatively stable all year. Although windows tend to be few, light enters through gaps in the stick walls and dimly illuminates the interior during the day. The tropical sun makes shade desirable, and the absence of windows prevents overheating. However, the lack of any insulation, the airy nature of the stick walls, and the close packing of buildings result in very little acoustic privacy.

- **Healthful:** The Kuna spend far less of their time indoors than typical people of industrialized countries. This reduces the need to be visually connected to the outdoors (e.g., with abundant windows) and allows their homes to be smaller. Their homes are well ventilated and free of VOCs, mold, and vapors from cleaning supplies. However, there can be pollutants resulting from cooking fires, improper sanitation, pets, and small livestock (such as pigs and chickens). Their latrines are built on piers over the water, dropping untreated waste directly into the sea.

- **Energy Efficient:** Although on some islands the Kuna have adopted gasoline generators and solar panels to produce electricity for lights and other modern conveniences, their consumption of energy remains minimal. This is enforced by the need to haul in all fossil fuels by hand, transporting them by canoes from the mainland.

- **Resource Efficient:** The Kuna sustainably harvest local materials from mainland property, transporting by boat everything used in the construction of their huts. This promotes efficient use of resources.

- **Long-Lived/Recyclable:** The wooden huts of the Kuna people satisfy the longevity principle by providing many years of shelter with minimal resource consumption. In addition, the huts can be recycled or biodegraded at the end of their useful life.

- **Minimal Environmental Impact:** Although the Kuna do not impact the environment significantly on a global scale (such as producing large quantities of carbon dioxide), they have drastically altered the ecology of their own islands. To make room for their densely packed shelters, they have removed most of the natural foliage and eliminated most of the natural animal life. However, their villages do not generally contribute to problems arising from storm water runoff, invasive plantings, chemical pollution, nighttime light pollution, and so forth.

The Kuna Village.  Photos courtesy of Scott Grinnell.

Chapter Summary

Green building promotes responsible and sustainable construction. Furthermore, green building creates spaces that are healthful, comfortable, and pleasant to occupy. Green buildings can improve occupant productivity and enhance entire communities, while simultaneously minimizing energy and resource consumption and environmental degradation. Across the world, many green-built structures have demonstrated that implementing green building practices does not necessarily cost more than conventional construction but creates structures that are less expensive to operate and maintain.

The United States constructs on average more than 1 million new buildings each year and provides energy and resources to operate and maintain almost 100 million existing buildings. The enormous demands of the building industry confront an ever-dwindling supply of natural resources and an uncertain source of energy. As a result, the manner in which buildings are constructed significantly impacts the well-being of society and the health of the environment.

Review Questions

1. Consider the six green building principles discussed in the text. Is any one of the principles more important than the others? If so, which one and why?

2. If two or more of the green building principles cannot be met simultaneously, what guiding philosophy should determine which principle to uphold?

3. How important are aesthetics in green buildings? How should aesthetics be balanced with or incorporated into the six green building principles?

4. What factors make green building practices more important today than ever before in history?

5. Studies show that people prefer to occupy rooms that offer views to the exterior on more than one side. Explain why this is so.

6. People tend to be comfortable over a range of temperatures that varies by only 6°F–8°F (such as 69°F–75°F). If people were comfortable over a much larger temperature range, such as 32°F–112°F for sedentary activities, how would this change the design of buildings?

7. Provide three examples of how a green building in a desert climate (e.g., New Mexico) could use less energy than a conventional building.
8. Which has a greater embodied energy, an Anasazi cliff dwelling or a village of teepees constructed from fir poles and buffalo skins? Explain your reasoning.
9. Consider the two structures described below. Does either satisfy the "long-lived" principle? If so, which would you say better satisfies the principle?
   a. A fishing shack built of driftwood and recycled metal roofing that is used for a single season and discarded.
   b. A traditional home built on the banks of the Guadalupe River that is severely damaged by flooding every 10–20 years.
10. Compare the building design for cold climates with that for hot, arid climates.

What similarities exist between the two? Explain why two dramatically different climates may produce similar building designs.
11. Why are many modern buildings constructed unsustainably when earlier structures were more sustainably built?
12. More than a thousand years ago, Norse explorers settled on the northern tip of Newfoundland and constructed wood-frame shelters covered by thick sod. They heated the buildings with open pit fires that vented smoke through small hatches in the ceiling. The shelters had no windows, had one small entrance, and were usually quite smoky. Consider the six principles of green building, and determine which principles were upheld and which were not. Overall, would you consider the sod shelters to be green buildings?

Endnotes

1 U.S. Census Bureau and U.S. Department of Housing and Urban Development. (2003). *New residential construction in February 2013.* Retrieved from http://www.census.gov/construction/nrc/
2 R.P. Leslie. (2003). Capturing the daylight dividend in buildings: Why and how? *Building and Environment*, 38(2), 381–385.
3 U.S. Environmental Protection Agency, Office of Air and Radiation. (1989). *Report to Congress on indoor air quality. Vol. 2, Assessment and control of indoor air pollution* (EPA 400-1-89-001C). Washington, DC: EPA, 1, 4–14.
4 U.S. Department of Labor. *OSHA technical manual.* Washington, DC: Author, Section III: Chapter 2, ASHRAE 62-1989 standard (TED 01-00-015, Effective Date: 1/20/1999).
5 U.S. Department of Energy, Center of Excellence for Sustainable Development, Smart Communities Network. (2010). *Green buildings introduction.* Retrieved from http://www.smartcommunities.ncat.org/buildings/gbintro.shtml
6 U.S. Census Bureau. (2010). *Median and average square feet of floor area in new single-family houses completed by location.* Retrieved from http://www.census.gov/const/C25Ann/sftotalmedavgsqft.pdf

Building Materials

Courtesy of Scott Grinnell.

Introduction

The selection and availability of construction materials strongly influence the design options for and the success of a building. Traditionally, local resources largely determined the choices for building materials. Today worldwide distribution makes available an enormous range of material options and empowers diverse design possibilities. Most building materials offer both advantages and limitations. The design and construction of a green building can be successful only with a clear understanding of the available materials and the environmental and social impacts associated with their manufacture and use. Careful selection of building materials is essential in green building and must include consideration of:

- The **embodied energy** of the product, which consists of the energy required to obtain, process, manufacture, and transport the material to the building site (see Table 2.1);
- **Resource depletion**, which measures the sustainability and renewability of the resource;
- **Environmental degradation**, which includes the erosion, pollution, and ecological destruction that occur during harvesting or mining;
- **Pollution** produced during manufacturing, processing, and transporting;
- **Toxicity**, which affects people and the environment at many stages: from the manufacturing process, to the finished product, to its final disposal; and
- **Performance and durability**, which determine the longevity and successful operation of the building.

Many construction materials share common features that influence their suitability for green building. These include:

- **Volatile organic compounds (VOCs)**: Adhesives, plastics, solvents, binding agents, and other building materials can outgas VOCs. The application of these products and the degree to which they outgas determine their overall impact. VOCs compromise not only the health of the building's occupants but also the health of the people manufacturing the products. In addition, some products release toxins into the environment after disposal, potentially impacting people and ecosystems unassociated with construction.
- **Portland cement**: Portland cement has a high embodied energy because its production entails pulverizing limestone and other materials and baking the mixture at a high temperature in a kiln. This energy-intensive process is usually achieved through the burning of fossil fuels, which releases many pollutants into the atmosphere, including sulfur dioxide, carbon dioxide, and mercury. Materials containing Portland cement range from concrete, mortar, stucco, and grout to fiber-cement siding and roofing. Despite the high embodied energy, products made from Portland cement tend to be very long-lived and can have valuable applications in green buildings.
- **Prefinished materials**: Engineered siding, laminate floor, prepainted metal roofing, and many other building materials can be finished or partially finished at the factory. While these materials generally have smaller health and disposal issues during construction, the environmental and health concerns typically occur at the manufacturing stage instead, affecting factory workers rather than construction workers.

TABLE 2.1 Embodied energy.

MATERIAL	BY WEIGHT (BTU/LB)	BY VOLUME (BTU/FT³)
Earthen Materials		
Adobe block	200	21,000
Cement	3400	420,000
Cement mortar	860	88,000
Concrete, ready mix	560	88,000
Dry wall plaster board	2600	160,000
Gravel, dry	43	4200
Rammed earth block	180	18,000
River rock	17	1800
Sand, dry	43	4400
Stone, local	340	52,000
Stone, imported	2900	450,000
Lumber & Sheathing		
Hardboard	10,000	370,000
Lumber, air dried and rough sawn	130	4600
Lumber, kiln dried and milled	1100	38,000
Lumber, glulam	2000	70,000
Medium density fiberboard (MDF)	5100	230,000
Oriented strand board (OSB)	3400	88,000
Plywood	4500	120,000
Roofing		
Asphalt	1500	200,000
Cement tiles	350	29,000
Cedar shingles	3800	4600
Slate shingles	340	58,000
Steel	14,000	6,900,000
Siding		
Aluminum, prepainted	94,000	16,000,000
Brick	1100	140,000
Engineered wood	5100	220,000
Fiber cement	3300	400,000
Steel, prepainted	15,000	7,300,000
Stucco	860	85,000
Vinyl	30,000	2,500,000
Wood, milled	1300	46,000
Wood, rough sawn	700	23,000
Flooring		
Carpet	31,132	Variable
Ceramic tile	1100	140,000
Linoleum	49,000	4,100,000
Stone, local	340	52,000
Stone, imported	2900	450,000
Vinyl	34,000	2,900,000
Wood	1300	38,000

(continued)

TABLE 2.1 Embodied energy. *(continued)*

MATERIAL	BY WEIGHT (BTU/LB)	BY VOLUME (BTU/FT³)
Insulation		
Cellulose	1400	3100
Expanded polystyrene	50,000	65,000
Extruded polystyrene	34,000	44,000
Fiberglass	13,000	27,000
Polyurethane foam	32,000	1,200,000
Straw, baled	100	840
Wool, recycled	6300	3800
Metals		
Aluminum	82,000	14,000,000
Aluminum, recycled	3500	600,000
Copper	30,000	17,000,000
Steel	14,000	6,900,000
Steel, recycled	4300	1,000,000
Miscellaneous		
Cotton fabric	61,000	550,000
Glass, window	6800	1,100,000
Paint, solvent-based	42,000	0.51/gal
Paint, water-based	38,000	0.46/gal
Paper	16,000	900,000
Paper, recycled	10,000	580,000
Plastic (ABS)	48,000	3,100,000
Polyester	23,000	210,000
Polypropylene	28,000	1,500,000
Polyurethane	32,000	1,200,000
Rubber, natural latex	29,000	1,700,000
Rubber, synthetic	47,000	4,600,000
Straw bale	100	840

© 2016 Cengage Learning®.

This chapter investigates a variety of common building materials, including treated lumber, composite boards, roofing, siding, flooring, insulation, and windows, and compares the advantages and limitations of the various products.

Treated Lumber

Treated lumber is wood infused with chemicals that are toxic to fungi and insects (Figure 2.1). To some degree, this process also makes the wood toxic to construction workers and building occupants. Nevertheless, treated lumber is used regularly in buildings for both interior and exterior applications. Despite the undesirability of adding toxicity to a building, treated lumber has several benefits. One of these benefits is the prolonged life of the wood, reducing the energy and resource requirements of frequent replacement. Another is the reduced demand for naturally decay-resistant species like cedar and redwood, which are slow growing and typically harvested unsustainably. Finally,

FIGURE 2.1 Treated lumber commonly includes copper as a fungicide, giving wood a greenish hue. Courtesy of Scott Grinnell.

treated lumber can come from fast-growing trees that would otherwise be unsuited for construction. Even with these benefits, green-built structures often employ designs that eliminate or reduce the need for treated lumber.

Composite Boards

Modern construction practices, particularly for residential buildings, regularly call for exterior wall sheathing for structural strength. Sheathing is also used for subfloors and as a substrate for roofing. It is generally some form of composite board, such as plywood, oriented strand board, or fiberboard.

Plywood is made from relatively large-diameter logs that are soaked in hot water to soften fibers and rotated on a lathe. Large blades peel off thin layers that are dried, coated in glue, cross-aligned, and squeezed together under pressure (Figure 2.2a). Most exterior-grade plywood uses phenol-formaldehyde glue, which is far more stable than the urea-formaldehyde glue used in specialty plywoods common in furniture and interior finishes. The relatively slow outgassing of phenol-formaldehyde glue reduces its toxicity and its likelihood of contributing to sick building syndrome. However, it is a known carcinogen and is not recommended for spaces with poor ventilation.

Oriented strand board (OSB) is produced by chipping logs into strands, cross-aligning the strands, spraying each layer with a binding agent, and squeezing the layers together under high pressure (Figure 2.2b). OSB typically uses either phenol-formaldehyde glue or isocyanate resin as a binding agent and, like plywood, slowly outgases. The manufacture of OSB is more resource efficient than that of plywood because it is able to utilize

FIGURE 2.2a Plywood consists of multiple layers of wood veneers glued together under high pressure. Courtesy of Scott Grinnell.

FIGURE 2.2b The manufacture of oriented strand board compresses together thin strands of shredded wood. The adhesive-coated strands are cross-aligned to improve its strength. Courtesy of Scott Grinnell.

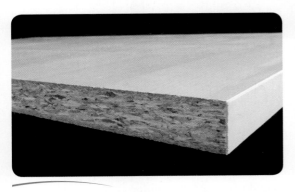

FIGURE 2.2c Medium-density fiberboard, often covered with a veneer of wood or vinyl, frequently replaces solid wood for the manufacture of cabinets, furniture, and doors. Courtesy of Scott Grinnell.

small-diameter logs and waste wood. Both plywood and OSB exhibit similar structural properties for construction, though plywood is generally better at resisting water damage and mold growth.

Fiberboard is created from pulpwood that is mechanically broken down into fibers, felted, and then reconstructed into sheets by heat and pressure (Figure 2.2c). Fiberboard products include **particleboard**, **medium-density fiberboard (MDF)**, and **hardboard**. Although hardboard is sometimes manufactured without a binding agent, particleboard and MDF usually contain urea-formaldehyde glue, which can contribute to poor indoor air quality. The manufacture of fiberboard is the most resource efficient of the three forms of sheathing, since it utilizes low-quality wood species.

Roofing

A roof must protect the building beneath it. It must endure scorching heat, battering rain, high winds, and, in some climates, hail, ice, and snow. Roofing material must be durable and capable of resisting the elements. Some roofs may be suitable for collecting rainwater; some may be fire resistant. Certain structures require lightweight roofing; for others, weight is not an issue. For all buildings, the material's embodied energy, resource depletion, and recyclability influence its appropriateness for green building.

A heavy roof requires stronger support structures beneath it, increasing the resource consumption and embodied energy of the substructure. Roofing that is recyclable or

TABLE 2.2   Comparison of roofing material.

MATERIAL	DURABILITY (YRS)	RAINWATER COLLECTION	FIRE RESISTANT	EMBODIED ENERGY	RECYCLABILITY	WEIGHT
Asphalt	15–30	No	No	High	No	Moderate
Cedar shakes	30–50	Yes	No	Low	Yes	Light
Clay tiles	50–80	Yes	Yes	High	Yes	Heavy
Fiber-cement tiles	20–50	Yes	Yes	High	No	Heavy
Metal	50+	Yes	Yes	High	Yes	Light
Slate tiles (local)	50–100	Yes	Yes	Low	Yes	Heavy
Sod/living	50+	Yes	Yes	Low	Yes	Very heavy

© 2016 Cengage Learning®.

that has been recycled reduces energy and resource consumption. Some types of roofing are best suited to particular climates; others, like asphalt, are used more widely.

This section describes common roofing materials found in the United States. The estimates of lifespan depend on installation, roof pitch, climate, prevailing weather, and maintenance and should be considered approximate. Table 2.2 compares the different types of roofing material described below.

Asphalt Shingles Composed of paper or fiberglass saturated in asphalt and top-coated with ceramic granules, asphalt shingles are the most common form of roofing in the United States, comprising two-thirds of the U.S. residential roofing market (Figure 2.3a). They are easy to install, are reliable, come in a variety of colors and patterns, and tend to be relatively inexpensive. However, regardless of color, all asphalt shingles have a dark substrate and can result in elevated rooftop temperatures. On sunny days, asphalt roofing can be 75°F–100°F hotter than the surrounding air temperature, increasing cooling demands in hot climates and contributing to an urban heat island effect. With a lifespan of only 15–30 years, asphalt shingles make up a significant portion of all building material waste, as they are rarely recycled. Toxic chemicals volatilize from asphalt in sunlight and leach from it in rain, making asphalt unsuitable for rain catchment systems.

Cedar Shakes Although wood roofing is natural, nontoxic, and recyclable, it has become illegal in many fire-prone areas due to its flammability. Cedar is a slow-growing tree that is rarely harvested sustainably. Roofs covered with cedar shakes have an expected lifespan of 30–50 years, are lightweight, and serve well for rain catchment systems (Figure 2.3b).

Clay Tiles Clay tiles are durable, fireproof, and recyclable but often have high embodied energy due to fossil fuels used in kiln baking. Clay is heavy and usually requires additional roof framing. Clay tile roofs have a lifespan of 50–80 years and are excellent for rain catchment (Figure 2.3c).

Fiber-Cement Tiles Composed of a fiber matrix embedded in Portland cement, fiber-cement tiles are durable, fireproof, lighter than clay tiles, and make good rain catchment roofs. They can be colored and textured in a variety of styles. Fiber-cement tiles have a life expectancy of 20–50 years, depending on the manufacture, and have a fairly high embodied energy.

FIGURE 2.3a Asphalt roofing is the most prevalent and widespread type of roofing in the United States. Courtesy of Scott Grinnell.

FIGURE 2.3b Cedar shake roofing grays as it weathers but outlasts most asphalt roofing. Courtesy of Scott Grinnell.

FIGURE 2.3c Clay tile roofing is durable but heavy. Courtesy of Scott Grinnell.

FIGURE 2.3d Standing-seam metal roofing is lightweight and sheds snow easily. Courtesy of Scott Grinnell.

FIGURE 2.3e Slate tile roofing comes in numerous shapes and colors. While heavy, it is extremely durable. Courtesy of Scott Grinnell.

FIGURE 2.3f People relied on traditional sod roofs for centuries before modern alternatives became available. This roof in Norway has lasted more than 100 years. Courtesy of Scott Grinnell.

Metal Roofing Metal roofing is available as shingles or in large sheets and can be made from steel, aluminum, or copper. It is durable, fireproof, and lightweight and can be recycled indefinitely. Metal roofing is excellent for rain catchment and has an expected lifespan of 50+ years. Even though metal roofing has a large embodied energy, it is durable and can be recycled at the end of its useful life (Figure 2.3d).

Slate Tiles While slate tiles are heavy, a slate roof is extremely durable, lasting 50–100 years or longer. Slate is natural stone that is quarried and split, and its environmental impact is usually quite small compared with other roofing products. However, slate is practical only if there is a local source, as the embodied energy of transporting it over distances can be large. Slate is dark and not recommended for hot, sunny climates. It is excellent for water collection systems (Figure 2.3e).

Sod Roof (Living Roof) A sod roof (living roof) consists of soil or some other growing medium established over a waterproof membrane. When saturated with water or covered in snow, living roofs can be quite heavy—even heavier than clay or slate and at least three times heavier than asphalt shingles. Therefore, living roofs require substantial additional structural support. However, living roofs can last for generations and offer many benefits; they:

- Mitigate storm water runoff by absorbing and slowly releasing moisture.
- Cool roof surfaces through the transpiration of plants, reducing energy requirements for cooling.
- Provide insulation that prevents overheating in the summer and heat loss in the winter.
- Provide wildlife habitat for birds, insects, and small animals.
- Improve air quality by absorbing pollutants and producing oxygen.
- Increase the longevity of existing roof membranes by blocking ultraviolet rays and preventing extreme surface temperatures.

When the soil and plants are obtained locally, living roofs have low embodied energy and resource depletion (Figure 2.3f).

Siding

Siding serves the same general purpose as roofing: it protects the building from outside elements. It is the first line of defense against the infiltration of water, wind, insects, and dust. Considerations for selecting siding are similar to those used for roofing: durability, embodied energy, resource depletion, fire resistance, weight, and recyclability. Table 2.3 shows a comparison of siding materials discussed below.

Brick and Stone When properly installed, brick and stone siding can last 50–100 years or more. Brick and stone are fireproof, recyclable, and durable. The embodied energy varies, depending on the material, manufacturing process, and source. Kiln-fired bricks have a large embodied energy, while the embodied energy of locally sourced stone can be quite low. Whereas most forms of siding adhere to exterior walls, brick and stone must be supported by the building foundation because of their extreme weight (Figure 2.4a).

TABLE 2.3 Comparison of siding materials.

MATERIAL	DURABILITY (YRS)	FIRE RESISTANT	EMBODIED ENERGY	RECYCLABILITY	WEIGHT
Brick and stone	50+	Yes	Variable	Yes	Heavy
Engineered wood	20–30	No	Moderate	No	Light
Fiber-cement	30–50	Yes	High	No	Heavy
Metal	30–50	Yes	High	Yes	Light
Stucco/plaster	50–60	Yes	High	No	Heavy
Vinyl	50	No	High	No	Light
Wood	50+	No	Low	Yes	Light

© 2016 Cengage Learning®.

Engineered Wood Engineered wood uses sawdust or wood flakes combined with an adhesive binding agent to make lightweight boards that look similar to wood. Engineered wood siding sometimes includes a tough waterproof surface layer that can be textured and colored in a variety of ways. The expected longevity depends on the manufacturing process, though most last 20–30 years. Engineered wood siding has low resource depletion and embodied energy due to its recycled content. However, the binding agents typically include VOCs, which may increase its toxicity (Figure 2.4b).

Fiber-Cement Durable and fireproof, fiber-cement siding is made from Portland cement, sand, and cellulose fiber. It can be colored and molded to look like almost anything and has a life expectancy of 30–50 years or more. It is quite heavy, however, and its attachment requires a solid substrate, making it more difficult to install than other types of siding. Fiber-cement siding has a large embodied energy but is long-lived (Figure 2.4c).

Metal Usually composed of aluminum or steel, metal siding is durable and fireproof and comes in a variety of colors and textures. Metal has a large embodied energy but can be recycled indefinitely. Selecting siding with recycled content reduces its embodied energy substantially. Steel can be expected to last 50 years, while aluminum has a life expectancy of 30–50 years.

Stucco/Plaster Made of Portland cement, lime, and sand, stucco/plaster is fireproof and largely maintenance-free. It can be tinted and textured in a variety of ways. Its life expectancy is 50–60 years, though protecting it from water can extend this. It has a relatively large embodied energy (Figure 2.4d).

Vinyl Made from polyvinyl chloride (PVC) plastic, vinyl siding is popular because it is relatively inexpensive and requires little maintenance. It does not rot or flake and can be made in virtually any color or texture. Although it never needs to be painted, vinyl siding can crack, fade, and grow dingy over time. In addition, the production of PVC produces toxic compounds that lead to a variety of health problems, including cancer and birth defects. Furthermore, PVC manufacturing wastes often contribute to groundwater contamination and other forms of pollution. Vinyl siding

FIGURE 2.4a Stone siding varies considerably among regions, depending on the available material. Courtesy of Scott Grinnell.

is lightweight and can be placed over existing siding, making it popular for remodeling. PVC fares poorly in fires and produces toxic smoke when burned. Vinyl siding has a life expectancy of 50 years, though it is rarely recycled at the end of its useful life (Figure 2.4e).

Wood   Solid wood siding is natural, renewable, and recyclable and can last 50–100 years or more when properly installed and maintained. When locally sourced, wood siding has low embodied energy. Unpreserved wood can degrade and in some climates presents unwanted opportunities for termites. Wood is commonly treated with stain or paint, which must be applied regularly and may contain toxins. Zero-VOC stains and paints should be used when feasible. Wood can be a poor choice in areas prone to fires (Figure 2.4f).

FIGURE 2.4b Engineered wood siding. Courtesy of Scott Grinnell.

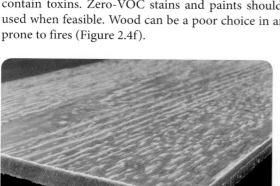

FIGURE 2.4c Fiber-cement siding can be manufactured to resemble stained wood. Courtesy of Scott Grinnell.

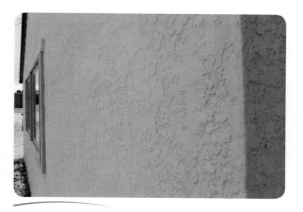

FIGURE 2.4d Stucco allows for a variety of colors and textures. Courtesy of Scott Grinnell.

FIGURE 2.4e Vinyl siding is inexpensive and easy to install. © Wendy Kaveny/Dreamstine.com.

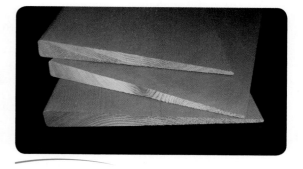

FIGURE 2.4f Solid wood siding is commonly stained or painted and is available in many different styles. This wedge-shaped cedar siding installs in horizontal, overlapping layers. Courtesy of Scott Grinnell.

Flooring

The type of interior flooring influences the character of the building and the well-being of its occupants. Flooring varies in smoothness, hardness, sound-deadening properties, insulation value, apparent warmth, and function. This section considers common types of flooring used in the United States. Table 2.4 shows a comparison of flooring materials discussed below.

Bamboo Bamboo flooring is composed of numerous strips of bamboo laminated together with a binding agent. Bamboo is a highly renewable, fast-growing grass common to China, Thailand, and other tropical regions. Although the energy associated with production is relatively modest, the long-distance shipping of bamboo flooring to North America increases its embodied energy substantially. It has a life expectancy of 30–50 years. Some bamboo flooring is available with nonformaldehyde glues, reducing possible health concerns (Figure 2.5a).

Carpets Carpets can be made from natural materials, such as jute, flax, hemp, cotton, wool, and animal hair, or from synthetic materials derived from petroleum products. Carpets made from natural materials tend to contain fewer toxins. Most synthetic carpets outgas VOCs, which can cause an array of health problems. Nonfireproof and potentially toxic when burned, carpets offer valuable insulation and sound-deadening properties and provide cushion to the floor. With a life expectancy of 10–20 years (depending on traffic), carpets tend to be rather short-lived compared to other flooring materials. The embodied energy of carpets varies, depending on material and process.

Ceramic Tile Created from natural clay, ceramic tile flooring provides a durable, nontoxic, fireproof, water-resistant, low-maintenance surface. Tile is heavy and offers thermal mass that can help moderate temperature swings. It has a rather large embodied energy due to fossil fuels used during high-temperature firing. Its life expectancy is 50 years or more when properly installed.

TABLE 2.4   Comparison of flooring materials.

MATERIAL	DURABILITY (YRS)	FIRE RESISTANT	EMBODIED ENERGY	RECYCLABILITY	OFF-GASSING
Bamboo	30–50	No	Moderate	Yes	Low
Natural carpets	10–20	No	Moderate	Yes	Low
Ceramic tile	50+	Yes	High	Yes	Very low
Concrete	50+	Yes	High	Yes	Very low
Cork	20–50	No	Moderate	Yes	Low
Engineered wood	30–50	No	Moderate	No	Moderate
Laminate	15–25	No	Moderate	No	High
Linoleum	25–50	No	Moderate	No	Low
Wood	50+	No	Low	Yes	Very low
Stone	100	Yes	Low	Yes	Very low
Vinyl	10–50	No	High	No	High

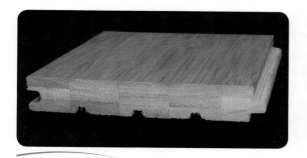

FIGURE 2.5a Bamboo flooring is made by laminating strips of bamboo into a solid board prior to milling the board into its final shape. The strips can be aligned either horizontally (as shown) or vertically. This type of tongue-and-groove flooring is nailed to the subfloor. Courtesy of Scott Grinnell.

FIGURE 2.5b Free-floating cork tiles snap together without the use of glue or nails and are an economical alternative to solid cork flooring. Layers of cork on the top and bottom surfaces sandwich a substrate of fiberboard. Courtesy of Scott Grinnell.

Concrete For buildings erected on concrete slabs, the slab itself can serve as the finished floor, avoiding the resource depletion and embodied energy associated with supplemental flooring. There are two methods that give concrete slabs a finished look. The first method adds an inert, nontoxic pigment to the concrete before it is poured. The pigmented concrete can then be smoothed or textured in a variety of ways. The second applies an acid to the surface after the concrete has cured. The acid reacts with the concrete, coloring the surface, but in the process, it produces noxious vapors. Concrete floors are long-lived, water resistant, and fireproof and have high thermal mass.

Cork Cork comes from the bark of the cork oak tree, which grows in the Mediterranean region. Cork bark can be removed without harm to the tree in a sustainable and renewable manner. The cork used in flooring is typically a waste product from the wine-stopper manufacturing industry. Ground cork is mixed with urethane binders and compressed under heat to form solid cork tiles that can be finished and refinished much like hardwood floors. Cork can also be cut into thin layers and adhered to fiberboard to make tongue-and-groove laminate flooring tiles, as shown in Figure 2.5b. As with any laminate flooring, the manufacturing process typically requires chemical adhesives such as formaldehyde. Cork is usually glued to the subfloor. Although nonfireproof, cork flooring offers many desirable properties. It is water resistant, sound deadening, shock absorbing, and hypoallergenic. It also provides thermal insulation. Its life expectancy varies, depending on the manufacturing process and installation. Solid cork flooring can last as long as solid wood flooring, making it very durable.

Engineered Wood Composed of three or more layers of wood veneer glued together, engineered wood flooring is thinner and more stable than solid wood, expanding and contracting less with changes in moisture and temperature. However, the off-gassing of glues and solvents used in its manufacture may pose health concerns. Engineered wood has a higher embodied energy than solid wood but smaller resource depletion, since it can be made of wood fragments. Its useful life is expected to be 30–50 years.

Laminate Composed of layers of paper, wood, and resin bonded together under heat and pressure, laminate flooring is often made to resemble wood. Like engineered wood, laminate flooring may contain VOCs, which pose health risks. Laminate floors are not fireproof and have an expected useful life of 15–25 years (Figure 2.5c).

Linoleum Composed of linseed oil, cork, wood flour, and powdered limestone, linoleum is a natural alternative to vinyl. Linoleum is often pressed onto a jute backing and adhered to the subfloor with zero-VOC glues. Although not fireproof, **linoleum flooring** is nontoxic and completely biodegradable. Furthermore, linseed oil is naturally antibacterial, making linoleum more sanitary than vinyl. Its production has lower embodied energy than vinyl, but at present, linoleum is shipped from Europe to North America, increasing its embodied energy considerably. Linoleum can be installed as tiles or sheets and has a life expectancy of 25–50 years (Figure 2.5d).

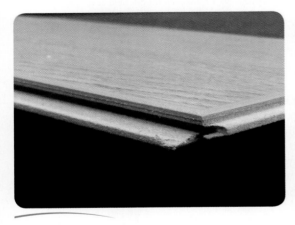

FIGURE 2.5c Although made to resemble traditional wood, laminate flooring contains no solid wood. Rather, it consists of a thin, wear-resistant layer over a substrate of fiberboard or resin-saturated paper. Laminate flooring snaps together without adhesives or nails. Courtesy of Scott Grinnell.

FIGURE 2.5d Linoleum flooring is available in sheets, as tiles that adhere to the subfloor with glue, or as free-floating snap-together tiles, as shown here. Courtesy of Scott Grinnell.

FIGURE 2.5e Solid wood flooring can either arrive prefinished, as shown, or be sanded and finished in place. Most prefinished wood flooring includes micro-bevels around the edges that leave small grooves between pieces. Courtesy of Scott Grinnell.

FIGURE 2.5f Quarried stone creates long-lasting floors, anchored to the subfloor with mortar and grout. Courtesy of Scott Grinnell.

Solid Wood Wood has been used for flooring for hundreds of years. It is a natural, re-newable, recyclable material that can last the lifetime of the building. Solid wood flooring has low embodied energy if sourced locally and can be repaired and refinished easily. Wood floors can come prefinished or be finished in place and are usually nailed to the subfloor (Figure 2.5e).

Stone Even more durable than tile, stone flooring shares all of the attributes of tile and may have low embodied energy if locally sourced. It is hard, heavy, fireproof, and water resistant. Its life expectancy is 100 years or more (Figure 2.5f).

Vinyl Vinyl flooring is popular because it is resilient, wear resistant, low maintenance, and easy to install. It comes in many colors and patterns and can be installed as tiles or large sheets. The production of vinyl flooring, like that of any PVC product, creates environmental concerns. It has a relatively large embodied energy, is not generally recycled, is not fireproof, and has a life expectancy of 10–50 years, depending on the quality and installation.

Heat Transfer and Insulation

In most climates, buildings must be insulated to maintain a comfortable interior temperature. The amount and type of insulation greatly influence the energy consumption of the building. The more extreme the climate, the more insulation is required. Only where the daily and seasonal temperatures allow buildings to remain comfortable without heating or cooling can insulation be ignored. This occurs predominantly in tropical climates.

Proper insulation is critical for green building because it is one of the primary methods of reducing energy consumption and creating a comfortable living space. Insulation prevents the unwanted transfer of heat, either from the warm interior of the building to a colder exterior or from a warmer exterior to the cooler interior of the building. There are three mechanisms of heat transfer, and different forms of insulation address each of these mechanisms with differing levels of success.

Heat Transfer

Heat always tends to move from a warmer place to a cooler place. For example, a hot cup of tea left at room temperature will gradually cool off as heat transfers *from* the tea *to* the room. Similarly, a cold glass of iced tea will warm up as heat transfers *from* the room *to* the iced tea. Considering the transfer of heat and the mechanisms responsible for it is essential when assessing the energy flows and insulation requirements of a building. Heat moves by three fundamental processes (see also Figure 2.6):

- Conduction is the transfer of heat by direct contact. For example, the handle of a spoon placed in a cup of hot tea will become warm as heat conducts from the tea along the spoon handle.
- Convection is the transfer of heat through the motion of a fluid, such as air or water. Warm air leaking through a gap in a door is replaced by cold drafts, transferring heat between the building and the outside. Similarly, the breeze created by a ceiling fan can move heat around a building.

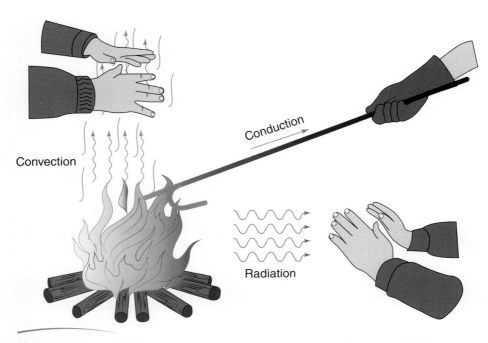

FIGURE 2.6 The three forms of heat transfer are conduction, convection, and radiation.
© 2016 Cengage Learning®.

- **Radiation** is the transfer of heat through electromagnetic waves, such as light. For instance, sunlight streaming through a window warms a building. Similarly, infrared radiation from the warm interior of a building can pass through a window at night, allowing heat to escape.

The effectiveness of insulation in preventing heat transfer depends on its composition, density, and method of installation. A measure of the resistance of a material to heat transfer by conduction is the **R-value**. The greater the R-value, the less heat passes through the material and, therefore, the greater the insulation level. Table 2.5 shows a comparison of R-values and other characteristics for the common types of insulation discussed below.

The R-value is the standard measurement for assessing the insulation value of materials sold in the United States. However, the R-value considers only *conductive heat transfer*. While conduction is a major form of heat transfer in buildings, significant air infiltration arising from major leaks around windows or gaps under doors may allow more heat to move by convection than through conduction. Therefore, in addition to being well insulated, buildings need to be tightly constructed to prevent unwanted air infiltration.

Insulation is available in several forms:

- **Loose-fill insulation** takes the form of granules or small fibrous pieces that can be poured, sprayed, or pumped into cavities, settling into irregularly shaped spaces. Commonly used in flat attics, loose-fill insulation offers effective self-leveling coverage. However, loose-fill insulation does not fully arrest air motion and, therefore, does not prevent convective heat loss.

TABLE 2.5	Comparison of insulation materials.			
TYPE	**FORMS**	**R-VALUE (PER INCH)**	**ADVANTAGES**	**DISADVANTAGES**
Cellulose	Loose fill	3.7 (at a density of 3.0 lb/ ft^3)	• High recycled content • Nontoxic • Renewable • Recyclable • Low embodied energy	• Boric acid is irritating to skin and respiratory tract • Absorbs water
Cementitious foam	Spray foam	3.9	• Seals gaps • Stops air movement • Fireproof • Nontoxic • Water proof • Contains no VOCs	• Moderately high embodied energy
Cotton	Batts	3.4	• High recycled content • Nontoxic • Renewable • Recyclable • Low embodied energy	• Boric acid is irritating to skin and respiratory tract • Absorbs water
Expanded poly-styrene (EPS)	Rigid board	4.2 (at a density of 2.0 lb/ft^3)	• High R-value • Recyclable • Stable insulation value	• Toxic smoke when burned • High embodied energy
Extruded poly-styrene (XPS)	Rigid board	5.0 (initially) 4.0 (stabilized value)	• High R-value	• Insulation properties decrease over time • Manufacturing process uses HCFCs • Toxic smoke when burned • High embodied energy
Fiberglass	Loose fill, rolls, and batts	3.1	• Plentiful resource • Fire resistant • Does not absorb water	• Irritating to skin and respiratory tract • May contain formaldehyde • Moderate embodied energy
Mineral wool	Loose fill, rolls, and batts	3.3	• High recycled content • Fire resistant • Low embodied energy • Does not absorb water	• Irritating to skin and respiratory tract
Plastic fiber	Batts	4.0	• High recycled content • Does not absorb water	• Toxic smoke when burned • May contain VOCs
Polyisocyanurate	Rigid board	7.0 (initially) 5.6 (stabilized)	• High R-value	• Insulation properties decrease over time • Manufacturing process uses HCFCs • Toxic smoke when burned • High embodied energy
Polyurethane	Spray foam	5.4	• High R-value • Seals gaps • Stops air movement	• Uses HCFCs • Toxic smoke when burned • High embodied energy

(continued)

TABLE 2.5   Comparison of insulation materials. *(continued)*

TYPE	FORMS	R-VALUE (PER INCH)	ADVANTAGES	DISADVANTAGES
Straw bales	Bales	1.8–2.0	• Low embodied energy • Nontoxic • Highly renewable • Recyclable	• Not fire resistant • Absorbs water
Vermiculite	Loose fill	2.4	• Fire resistant	• May contain asbestos • Lower R-value
Wool	Batts	3.5	• Nontoxic • Renewable • Recyclable • Low embodied energy	• Boric acid is irritating to skin and respiratory tract

© 2016 Cengage Learning®.

- **Batt or roll insulation** resembles flexible, blanketlike slabs of matted material that fit tightly into cavities when cut and properly installed. Batt or roll insulation is often used in wall cavities, and its effectiveness in preventing air motion depends on the material and care of installation.
- **Rigid foam insulation** is commonly available in 4-foot by 8-foot sheets and works well beneath concrete slabs, when spanning walls and ceilings, and when cut to fit within cavities. Rigid sheets provide an effective air barrier when seams are sealed.
- **Spray-in-place foam insulation** usually emerges from a nozzle as an aerated liquid that rapidly expands to fill and seal a cavity. Spray-in-place foams have the advantage of minimizing conductive heat transfer through high R-values, while simultaneously arresting convective heat transfer by thoroughly sealing penetrations. Most spray-in-place foams are petroleum-based products with high embodied energy.

Common Types of Insulation

Most forms of insulation have advantages and disadvantages that need to be considered prior to designing a building. Some forms of insulation come from natural materials, while others are petroleum based. Some are made of recycled products, such as newspapers or denim. Some allow the movement of air and water vapor, while others act as vapor barriers. Some types of insulation have low embodied energy and resource depletion, while others have quite high values.

Forms of insulation that are flammable or able to absorb and store moisture are commonly treated with boric acid as a fire retardant, mold inhibitor, and insect repellent. Boric acid is nontoxic but can be corrosive to metal and irritating to the skin, eyes, and respiratory tract of the installers.

The production of rigid foam insulation requires some kind of blowing agent during manufacturing that aerates the foam, producing microscopic bubbles that increase the insulation value. **HCFCs (hydrochlorofluorocarbons)** became commonplace as blowing agents after chlorofluorocarbons (CFCs) were globally banned due to ozone depletion.

While HCFCs are considerably better for the environment than CFCs, they contribute to climate change and still pose harm to the ozone layer. International agreements have scheduled the elimination of HCFCs by 2030. The environmental impact of these blowing agents is an important consideration when selecting appropriate insulation for green building.

Cellulose insulation is made from recycled paper (primarily newsprint) that is shredded into small, fibrous pieces. These pieces pack closely together, inhibiting airflow more effectively than fiberglass or mineral wool, and provide an insulation value of R-3.7 per inch (at a density of 3.0 lb/ft^3). Cellulose has the ability to absorb and store water and is usually treated with boric acid. It is nontoxic and has very low embodied energy (Figure 2.7a).

Cementitious foam insulation is magnesium oxide blown with air and water to create porous cement. It can be a spray-in-place foam that expands to seal and fill cavities or come as precast sheet insulation. The trapped air pockets produce an estimated insulating value of R-3.9 per inch. It contains no VOCs and is durable, fireproof, nontoxic, resistant to water and mold, and effective at deadening noise. Furthermore, it contains only 20%–40% the embodied energy of Portland cement.

Cotton insulation consists of 85% recycled cotton (such as blue jean manufacturing trim waste) and 15% plastic fibers. It is usually formed into batts and treated with boric acid. Like cellulose, cotton insulation can absorb and retain water. It is nontoxic, renewable, and recyclable and has an insulation value of R-3.4 per inch (Figure 2.7b).

Expanded polystyrene (EPS) insulation, sometimes called "bead board," is a closed-cell foam made from polystyrene beads expanded and fused together. It is naturally white in color and can be molded into various forms for packaging or made into rigid sheets for construction (Figure 2.7c). EPS uses pentane as a blowing agent, most of which is

FIGURE 2.7a Cellulose insulation consists of pulverized recycled paper and a powdered flame retardant, such as boric acid. Courtesy of Scott Grinnell.

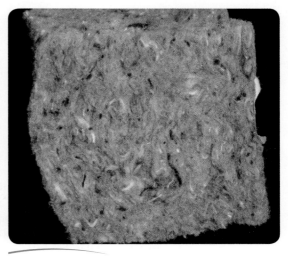

FIGURE 2.7b Cotton insulation comes from manufacturing waste, including trimmings from blue jeans. Courtesy of Scott Grinnell.

FIGURE 2.7c Expanded polystyrene insulation is made of closely packed polystyrene beads. Courtesy of Scott Grinnell.

FIGURE 2.7d The extruding process of XPS creates a finer-grained structure that is more visibly uniform than EPS. The process also enables companies to add various colorants. Courtesy of Scott Grinnell.

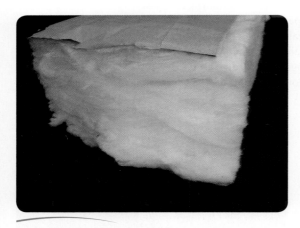

FIGURE 2.7e Fiberglass insulation can be manufactured without formaldehyde binders, as shown here. The paper backing acts as a vapor retarder that also limits convection. Courtesy of Scott Grinnell.

recaptured at the manufacturing plant, and has a stable insulation value of R-4.2 per inch (at a density of 2.0 lb/ft^3). EPS has a lower overall environmental impact than many other types of rigid foam because it is more easily recycled and does not contain HCFCs. EPS is flammable and produces thick smoke when burned.

Extruded polystyrene (XPS) insulation is also a closed-cell foam made from polystyrene beads. In the manufacture of XPS, the beads are melted and extruded using a hydrocarbon gas (like an HCFC) or carbon dioxide as a blowing agent. XPS can be formed into rigid boards that are usually colored by the manufacturer to designate brand (Figure 2.7d). Although XPS begins with an insulation value of R-5.0, this drops to R-4.0 after several years as the blowing agent gradually escapes and is replaced by air. Like EPS, XPS will burn when exposed to flame, producing thick smoke.

Fiberglass insulation is the most commonly used form of insulation in the United States. It is made from molten glass that is spun or blown into fine fibers and is available as loose fill, rolls, or batts (Figure 2.7e). It is fire resistant and has an insulation value of R-3.1 per inch. Fiberglass can be manufactured without binding resins and as a result is nontoxic, though the fine glass fibers can cause irritation to the skin, eyes, and respiratory tract during installation. Some brands of fiberglass contain formaldehyde, which can result in off-gassing. Fiberglass is permeable to air and water vapor and will perform below its rated R-value if its installation does not prevent convective heat transfer.

Mineral wool insulation is made from blast furnace slag or natural minerals like basalt and spun into fine fibers, much like fiberglass (Figure 2.7f). It contains roughly 75% postindustrial recycled content and is inherently fire resistant. Like fiberglass, it is available as loose fill, rolls, or batts and has a similar (or slightly higher) insulation value. It tends to be less irritating to the skin, eyes, and respiratory tract than fiberglass. Mineral wool generally has a lower embodied energy than fiberglass due to its recycled content. Like fiberglass, mineral wool is permeable to air and water vapor and will perform below its rated R-value if air movement is not prevented.

Plastic fiber insulation is made from polyethylene terephthalate (PET) plastic (such as recycled plastic milk bottles) and formed into batts similar to fiberglass. The insulation is treated with a fire retardant so it does not readily burn; however, it does melt when exposed to flame. It insulates at R-4.0 per inch (at a density of 3.0 lb/ft^3). Although derived from petroleum, its embodied energy is considered low due to its recycled nature. Like fiberglass and mineral wool, it is permeable to air and water vapor and will perform below its rated R-value if its installation allows convective heat transfer.

Polyisocyanurate insulation is a closed-cell rigid foam board that uses a low-conductivity gas (usually an HCFC) to fill the cells. This low-conductivity gas gives the foam excellent insulation properties (R-7.0 per inch). However, over time, some of the gas escapes and is gradually replaced by air, which has a lesser insulating value. Consequently, the insulating value of polyisocyanurate decreases with time, becoming about R-5.6 per inch after several years. Polyisocyanurate is a petroleum-based product with a high embodied energy and will produce toxic smoke when burned.

Polyurethane insulation is a closed-cell foam that also uses an HCFC as the blowing agent (Figure 2.7g). Its environmental impact and insulation performance are similar to those of polyisocyanurate. Polyurethane is available as a spray-in-place foam that rapidly expands to seal and fill cavities, making it particularly useful in preventing heat loss through convection.

FIGURE 2.7f Mineral wool is generally less irritating than fiberglass during installation and may contain a higher recycled content. Courtesy of Scott Grinnell.

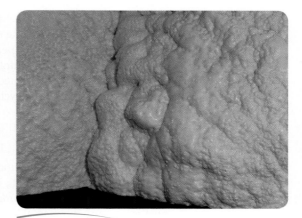

FIGURE 2.7g Spray-in-place foam insulation, such as the polyurethane insulation shown here, fills cavities and stops convective heat loss in addition to minimizing conduction. Courtesy of Scott Grinnell.

FIGURE 2.7h Straw bales must be kept dry, such as with an elevated sill as shown. Courtesy of Scott Grinnell.

FIGURE 2.7i Vermiculite insulation consists of lightweight mineral pellets. It is rarely used any more due to possible asbestos contamination. Courtesy of Scott Grinnell.

Straw bales have an insulation value between R-2.4 to R-3.0 per inch, but straw-bale construction typically yields a lower value (e.g., R-2.0 or less) due to air movement between the bales. Straw is a highly renewable agricultural waste product with very low embodied energy and is naturally nontoxic. Straw-bale insulation usually uses untreated straw bales and, therefore, requires a design that keeps the bales dry (Figure 2.7h). Straw will smolder when exposed to flame and must be sealed behind drywall, stucco, or other fire resistant material.

Vermiculite insulation is a form of loose-fill insulation made by rapidly heating silicate minerals. Commonly found in homes built before 1950, it is no longer manufactured because some sources of vermiculite were found to contain traces of asbestos—a known human carcinogen. Vermiculite consists of small, lightweight pellets and has an insulation value of about R-2.4 per inch (Figure 2.7i).

Wool (from sheep) can be treated with boric acid and used as insulation, providing an insulation value of R-3.5 per inch. Wool insulation has the advantage of being able to hold relatively large amounts of water and then dry out again without damage to the building. However, repeated wetting and drying can leach away boric acid. Wool is nontoxic, renewable, and recyclable and has low embodied energy.

Windows

Windows are an essential element of any building suitable for human occupancy. They provide light, ventilation, and a connection to the outside world. They are also one of the most poorly insulated components of a building, typically accounting for a significant portion of a structure's total heat transfer. In very cold or hot climates, large windows can severely impact the comfort of a building and radically increase its energy requirements. While a well-insulated wall may have an R-value between R-20 to R-40, even the best windows have insulation values of only R-4 to R-8, and most are less than R-3.

Appropriate window selection and placement depend on the climate, prevailing weather, position of the window within the building, and intended purpose. Windows used in cold climate buildings should be small, few in number, and placed chiefly on the sunward side, where solar heating during the day can compensate for heat loss. In hot climates, windows should be shielded from direct sunlight or placed predominantly on the shady side of the building.

Common Window Types

Windows that open to allow ventilation either pivot about hinges or permit sashes to slide past each other. Hinged windows can pivot into the living area or toward the outside; in either case, they require space for their movement. Sliding windows have the advantage of not requiring this space. However, hinged windows can be pulled tight against elastic weather-stripping, called **compression seals**, and are more effective at preventing air infiltration than windows that slide. The movement of air through gaps within and around windows is a major source of convective heat transfer that compromises energy efficiency. Figure 2.8 shows examples of common window types described below.

Awning windows (Figure 2.8a) pivot about hinges and open upward to the outside. Their range of motion is comparatively small, but they offer the advantage of providing ventilation during rain without moisture entering the building. They close tightly, using compression seals, and minimize air infiltration. Screens are set on the inside of the window, allowing accumulated dust to fall into the living space.

Casement windows (Figure 2.8b) are hinged on one side and crank or push open, offering excellent ventilation. When opened fully, however, the weight bearing on the hinges can be substantial, limiting the size of casement windows. Furthermore, casements have the disadvantage of exposing the interior surface of the window to the elements when open. This can result in water spots and sun damage and may require refinishing more often. Casement windows close tightly, using compression seals, and minimize air infiltration. Screens are set on the inside of the window, allowing accumulated dust to enter the living space.

Double-hung windows (Figure 2.8c) permit individual sashes to slide up and down past each other. The ability to open the top sash independently from the bottom allows ventilation to occur at the top, at the bottom, or from some of each. While good for ventilation and light, these classic windows do not seal well, compromising energy efficiency. Screens are set on the outside, keeping dust outside.

Gliding windows (Figure 2.8d) allow one or more sashes to slide past each other horizontally. This common style is easy to open, can be made in large sizes, and is less obstructive than pivoting windows. However, like double-hung windows, gliding windows do not seal well, compromising energy efficiency. Screens are set on the outside, keeping dust outside.

Hopper windows (Figure 2.8e) are hinged much like awning windows but open into the living space rather than to the outside. The screens are on the outside, keeping dust outside. Depending on their method of operation, hopper windows may also be able to prevent rain from entering the living space.

Picture windows (Figure 2.8f) are nonopening windows that provide light but not ventilation. Because they are not limited by moving parts, picture windows can be made in almost any size and shape.

(a) Awning

(b) Casement

(c) Double-hung

(d) Gliding

(e) Hopper

(f) Picture

FIGURE 2.8 Each of these six common window types offers different benefits and limitations. The choice of window for a particular situation usually depends on a variety of factors, including personal preference.
© 2016 Cengage Learning®.

Window Performance

Window performance measures the ability of a window to provide light and ventilation, while minimizing unwanted heat transfer. Factors that influence window performance include the number of panes, the quality of the frame, selective coatings, the type of seals (for operable windows), and the manner and care of installation (Figure 2.9).

- **Multiple panes**: Windows with more than one pane provide better insulation than single-pane windows. An inert gas, such as argon, commonly fills spaces between panes. The smaller the space separating the panes, the more heat is transmitted by conduction as heat moves through the gas; the larger the space, the more heat is transmitted by convection as circulations between panes develop. The ideal spacing between panes depends on the window design but is usually between 5/8 inch and 3/4 inch.

- **Frame quality**: Heat travels not only through window glass but also through the frame. A **thermal break**—any nonconductive material that separates the inside of a window frame from the outside and reduces conductive heat loss—is essential to the performance of an insulated window. Without a thermal break (which might be a strip of rubber or urethane), an aluminum frame can conduct more heat than the entire surface area of glass in a large window. Thermal breaks not only save energy but also prevent undue condensation. Excessive condensation can damage frames and sills and lead to mold and mildew growth.

- **Selective coatings**: Glass can be coated with different types of transparent films to selectively block particular wavelengths. Most new windows include a **low-E (low-emissivity) coating** that reduces the transmission of infrared radiation. Low-E coatings allow the passage of visible sunlight but prevent heat from radiating back out, an effective means of trapping solar energy. Other coatings reduce the passage of ultraviolet light and minimize fading and bleaching. Glass can also be tinted—principally gray, bronze, or green—as a means of reflecting a portion of the light. **Tinted glass** is usually used to reduce solar heat gain or for purposes of privacy.

- **Type of seal**: Compression seals found in awning, hopper, and casement windows let in as little as $1/10$ the air infiltration admitted by sliding seals found in double-hung and gliding windows. For windows of poor quality, air infiltration often comprises the dominant form of heat transfer.

- **Quality of installation**: Air infiltration occurs not only around seals within the windows but also in gaps between window units and the rough framing of the

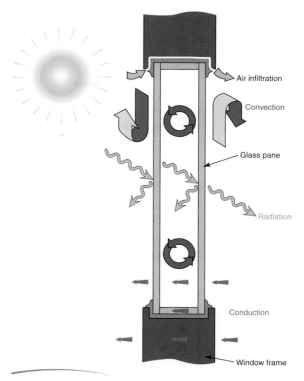

FIGURE 2.9 Windows often dominate a building's total heat transfer. Convection occurs from two sources: (1) the infiltration of air around window frames and through gaps in movable parts and (2) the natural movement of air on either side of the window and within spaces between panes. Radiation passes through windows predominantly in the form of visible light or infrared energy. Conduction occurs through the window frame and between panes of glass. All three forms contribute to the overall heat transfer. © 2016 Cengage Learning®.

building. During construction, these gaps need to be sealed and well insulated, such as with spray-in-place foams. Window performance is determined not only by the construction and components of the window itself but also by the quality of installation.

Four common parameters measure the performance of windows:

- The **U-factor** measures the heat that the window loses by conduction, where U-factor = 1/R-value. Hence, a U-factor of 0.25 represents an R-value of 4.0. The lower the U-value, the better the insulation value.
- The **solar heat gain coefficient (SHGC)** is the fraction of the sun's heat that passes through the window and is given as a number between 0 and 1. In hot climates, a low SHGC is generally desired, while a high SHGC is usually selected in cold climates.
- **Visible transmittance (VT)** is the amount of visible light that passes through the window and is given as a percentage between 0% and 100%. The VT determines the brightness and ambience of a room. It can also influence the health of indoor plants and the fading of artwork, carpets, and furniture.
- **Air infiltration** is the amount of air that leaks around the seals of a window and is measured in cubic feet per minute per square inch of window surface (cfm/in^2). The smaller the value, the less air infiltration and the better the insulation value of the window.

CASE STUDY 2.1

Stick-Built Home

Location: Northern Wisconsin (46.5° N latitude)

Year built: 2011

Size: 1750 ft^2

Cost: $166,000 ($95/ft^2)

Description: This two-story, four-bedroom home was constructed using standard dimensional lumber and conventional framing techniques. However, the application of green building practices through all phases of the construction process created an unconventional structure. Careful design and project management minimized resource consumption, greatly reduced waste, and created a comfortable, healthful, and energy-efficient structure.

The construction of a home involves numerous stages, each of which depends on the choice of building materials. These stages include preparing the site, laying the foundation, framing the building envelope, inserting windows and doors, adding roofing and siding, installing the internal systems (electrical, plumbing, and HVAC), applying appropriate insulation, finishing the interior (drywall, flooring, trim, cabinets, and so forth), and managing waste. Green building requires careful attention at all phases of construction.

Site preparation entails removing trees and other vegetation, leveling the ground, and digging footings. In this case, the building site required the removal of several large oak trees. Instead of cutting the trees into firewood, the logs were sectioned into boards on site and milled locally, becoming treads for the staircase and solid wood flooring for the second story.

The **foundation** supports the building and must be solid and durable. The foundation for this home consists of a pigmented concrete slab that not only establishes stability for the building but also serves as the finished floor for the ground level. This feature reduces resource consumption by avoiding supplementary floor material.

Framing the **building envelope** in an efficient manner requires meticulous attention to detail. Conventional stick-built structures consume more wood

Standard dimensional lumber and plywood give form to this structure. Courtesy of Scott Grinnell.

than structurally necessary, wasting materials and reducing the structure's insulation value, since solid wood provides comparatively poor insulation. The design of this building positioned exterior wall studs directly in line with floor and roof trusses, bearing the load of the building with minimal lumber. This permitted a uniform stud spacing of 24 inches, reducing material use by 33% over the standard 16-inch stud spacing. Careful placement of windows and doors upheld the 24-inch positioning without adding extra studs or massive headers. Engineered trusses for the roof and second-story floor made use of small-dimension solid wood lumber to create strong structures with local and sustainably harvested timber. The trusses contain no adhesives or other sources of VOCs that could compromise indoor air quality. The layout of the rooms and the dimensions of the structure took into account standardized building components like plywood, drywall, and lumber to minimize waste. Meticulous sealing of all seams, gaps, joints, and penetrations throughout the exterior walls produced a very tight envelope.

Windows and doors can significantly compromise a building's energy efficiency. To minimize energy loss, the owners selected triple-pane awning-style windows with multiple low-E coatings. The three panes of glass, separated by argon, reduce conduction and provide a U-factor of 0.24. The awning-style construction incorporates tight-fitting compression seals, inhibiting air infiltration. The low-E coatings greatly reduce the passage of infrared radiation. Careful sealing around window frames with expanding polyurethane foam prevents air infiltration around the window units. Despite these efforts, the windows remain the most poorly insulated component of the building envelope, providing only R-4.2. Consequently, the owners limited the number and size of windows and placed most

Two-inch-thick EPS rigid foam adds insulation to the exterior of the structure, subsequently protected by wood siding. Courtesy of Scott Grinnell.

of them on the south side of the building, where winter sunlight could warm interior spaces and compensate for heat loss. The building design includes an entry room that isolates the main door from the living spaces. The entry room buffers air flow between the interior and exterior and restricts unwanted heat transfer.

Roofing and siding must offer durable protection for the building. The design of this house includes two roofs: a primary steel roof above the second story and a sod roof above the protruding section of the first floor. The light-colored steel roof reflects summer light, keeping the building cool, and promises long-lasting protection. The sod roof provides insulation for the first floor and cool fresh air for the second. Additionally, the sod roof minimizes summer glare and enhances a visual connection to the outside. Generous 30-inch roof eaves protect the building's solid wood siding from weather, and buried drain tile at the eaves' drip edge conveys water to storage tanks for garden irrigation.

Internal systems for the house include a high-efficiency **energy recovery ventilator (ERV)** that captures and exchanges up to 70% of the energy between stale exhausted air from the interior and fresh make-up air from the outside. All tightly constructed buildings require suitable ventilation to maintain high indoor air quality, and the ERV of this home provides this ventilation, while preserving much of the energy. Energy-efficient ceiling fans cool the house during the summer, greatly reducing electrical demands over conventional air conditioning units. Polyethylene tubing embedded in the concrete slab of the lower level and in sand beneath the wood floor of the second story circulates warm water from a ground-sourced heat pump, creating an efficient and comfortable heat supply. Energy-efficient appliances complete the internal systems.

Insulation is essential for energy-efficient buildings, particularly in cold climates. For this home, densely packed cellulose insulation fills the 2- × 6-inch wall cavities, supplemented by 2 inches of EPS rigid foam outside the plywood sheathing, giving the walls an insulation value of R-30. The rigid foam placed on the outside of the studs acts as a thermal break for the entire house, reducing conduction through the wood framing. External connections, such as electrical outlets, mount on the siding rather than insetting into the foam, to avoid compromising the insulation value. Furthermore, utilities enter the house through underground conduits rather than directly through the walls. Densely packed cellulose also insulates the second-story ceiling, providing an insulation value greater than R-60. Four inches of extruded polystyrene rigid

Densely packed cellulose insulation fills the interior wall cavities, pumped through holes in a fabric mesh. Courtesy of Scott Grinnell.

foam completely surround the concrete slab, insulating the floor from the ground with a value of R-17 or greater.

Interior surfaces consist entirely of nonsynthetic materials, including a mixture of solid wood paneling and gypsum board for walls and ceilings, solid oak boards for the upstairs floors (harvested from the building site), ceramic tile for the floor of the upstairs bathroom, and solid wood for countertops, bookcases, furniture, and trim. Zero-VOC paints brighten the walls and ceilings throughout the house.

Waste management and minimization are an important aspect of green building. For this home, the materials not recycled or reused on site consisted primarily of drywall scraps and plastic packaging from cellulose bundles. These amounted to two loads in a pick-up truck, which went to the landfill. The project required no dumpster and released no toxic compounds into the atmosphere. Each day after construction, the owners collected and removed all wood scraps, bent nails, dropped screws, and so forth, keeping the building site clean and preventing debris from entering the surrounding environment.

Additionally, the owners chose to purchase all materials from local suppliers and used locally harvested and manufactured materials. This supported the regional community and minimized the embodied energy of transportation.

CASE STUDY 2.2

Timber Frame Home with Structural Insulated Panels

Location: Central Wisconsin (44.5° N latitude)

Year built: 2001

Size: 2250 ft^2

Cost: $250,000 ($111/ft^2)

Description: This two-story, three-bedroom home incorporates two nonconventional building techniques: timber frame construction and structural insulated panels:

- Timber frame construction replaces standard dimensional lumber with large timbers custom cut to fit together with wooden dowels or pegs. At the factory, the manufacturers shape and label each timber according to the design and then ship the ready-to-assemble pieces to the construction site. Workers assemble the pieces on the ground, and cranes lift the completed sections into place. Although generally more expensive than conventional construction, timber frame structures are extremely durable, can be assembled very quickly, and offer an attractive aesthetic style.

- Structural insulated panels (SIPs) are made by sandwiching rigid foam between two sheets of oriented strand board (or other composite board). Adhesives bond the materials together under high pressure to create a structural unit that often outperforms conventional construction in strength, stability, and insulation value. SIP construction does away with framing studs and other thermal breaks and minimizes air infiltration caused

by seams, gaps, and voids, thereby eliminating cold spots and providing better overall insulation. SIPs also provide better sound-deadening performance than most conventional construction. Although SIPs are made of flammable material, the absence of air within the panels decreases the spread of flames and makes SIP construction no more dangerous during a fire than conventional construction. SIPs are custom-made at the factory and typically require a crane for installation. When building with SIPs, the locations of pipes, wires, and outlets need to be determined precisely so they can be precut at the factory. SIP construction generally involves a higher embodied energy than conventional techniques because of the high foam content.

Solid oak timbers frame this structure, held together by notches and pegs. Courtesy of Brent and Amy Wiersma.

SIPs complement timber frame construction by rapidly completing the building envelope. In this case, workers assembled the timber frame structure in a single day and installed the structural insulated panels in two additional days. Within five days, the workers had installed the roof and

(a)   (b)

Structural insulated panels, lowered into place with a crane, minimize labor and assemble quickly. Courtesy of Brent and Amy Wiersma.

made the building weather tight. Conventional construction often requires a month or more to achieve this level of progress.

Built into a south-facing embankment, the lower level of this two-story house is largely underground, exposed primarily on the south side. This feature reduces exposure and conserves energy. Site preparation included excavating the lower level and hauling in gravel to facilitate drainage around the structure. The foundation consists of poured concrete, which also forms the three subterranean walls. Conventional 2- × 6-inch stud construction surrounds the concrete walls on the interior and creates the fourth wall on the exposed south side. Fiberglass insulation fills the stud cavities, and 2 inches of XPS rigid foam protect the outside of the concrete, creating a total insulation value of R-28 for the lower level. Engineered floor trusses span the concrete walls, creating the floor for the upper level.

The SIP construction for the upstairs provides a continuous insulation value of R-24 for the walls and R-38 for the ceiling. The owners selected triple-pane casement-style windows with multiple low-E coatings, offering an overall U-factor of 0.24. The casement windows offer excellent ventilation and close tightly to prevent unwanted air infiltration.

A light-colored steel roof reflects summer heat, sheds snow easily in the winter, and will last for generations. Vinyl siding shields the exterior of the SIPs from rain and weather, and stucco protects the exposed portion of the lower level.

The internal systems of the house include a high-efficiency **heat recovery ventilator (HRV)** to provide continuous fresh air, while capturing and exchanging energy. Ceiling fans provide summer cooling, and a propane boiler supplies heat by circulating hot water through pipes in the concrete slab of the lower level. A wood stove on the upper level supplements the in-floor system.

A combination of gypsum board and solid wood paneling finishes off the interior walls and ceilings. The floors combine a mixture of synthetic carpet and ceramic tile.

A standing-seam metal roof covers the SIP roof panels.
Courtesy of Brent and Amy Wiersma.

CASE STUDY 2.3

Insulated Concrete Form Home

Location: Wisconsin (43° N latitude)

Year built: 2004

Size: 4080 ft² main (6730 ft² including finished basement)

Cost: $800,000 ($120/ft²)

Description: This two-story, five-bedroom home with a full finished basement incorporates another common alternative to conventional stick frame construction: the use of insulated concrete forms.

Insulated concrete forms provide insulation, as well as thermal mass, when concrete fills the central cavity. Courtesy of Scott Grinnell.

Insulated concrete forms (ICFs) are interlocking foam blocks that can be stacked and filled with concrete. These forms remain in place after construction and provide permanent insulation for the building. They can serve as walls, floors, or roofs and are usually composed of EPS or XPS foam reinforced with plastic spanners. Compared to conventional construction, ICFs typically yield greater strength and stability, better sound-deadening properties, and higher insulation values. Although both concrete and foam have high embodied energy, the benefits of ICFs make them popular for green building.

This home uses ICFs for the exterior walls and relies on conventional framing for the roof and interior partitions. The ICF walls consist of 2-inch-thick EPS foam on both sides of a 6-inch reinforced concrete core, providing an insulation value of R-17. Unlike conventional construction, however, ICFs prevent nearly all air infiltration, and the high mass of the concrete core creates a more stable interior temperature, which reduces heating and cooling demands. The high-mass ICF walls demand an exceptionally stable foundation, and site preparation for this project included soil compaction and a base of crushed rock to minimize settling. In addition, drain tile embedded in crushed rock surrounds the basement walls and diverts water away from the building.

Careful installation and sealing around all windows and doors complete the tight envelope of this building. Double-paned, low-E, casement-style windows offer very low air infiltration and a U-factor of 0.30.

ICF forms must be shielded from weather and sources of ignition. This house uses gypsum board on the interior and stucco and stone siding to protect the exterior. Asphalt shingles seal the numerous dormers and valleys of the roof.

The internal systems include an HRV to provide fresh air, while reducing heat loss; an in-floor hot water heating system for the basement and garage; and independent high-efficiency forced-air furnaces for each of the main floors. A high-efficiency gas fireplace in the master bedroom and a sealed-combustion wood stove in the family room supplement the primary heating systems. Both forced-air furnaces include electronic air cleaners and ultraviolet lamps to remove dust, smoke, pet dander, and allergens and to kill bacteria and other microbes.

Each of the main floors also includes independent air conditioning units. As with the furnaces, the independent systems increase energy efficiency by conditioning each floor separately, depending on the needs of the occupants.

Cellulose fiber fills the attic spaces and provides a ceiling insulation value of R-50 or greater, while rigid foam insulates the basement slab to R-10. The house uses a mixture of flooring types—predominantly porcelain tile and solid hardwood, with synthetic carpet finishing off some of the rooms.

The owners employed low-VOC paints, purchased sustainably harvested lumber, and minimized waste by reusing and recycling most of the construction materials. Due to the many unrecyclable scraps from the insulated concrete forms, however, the owners could not eliminate the need for a dumpster.

The finished home uses stucco and stone siding to protect the ICF exterior. Courtesy of Scott Grinnell.

Further Learning CALCULATING HEAT TRANSFER BY CONDUCTION

Of the three forms of heat transfer, conduction is generally the simplest to determine when assessing building performance. Tight construction with well-sealed windows and doors minimizes convective heat transfer. Selective window coatings and reflective barriers reduce radiative heat transfer. To diminish conductive heat transfer, high levels of insulation are important.

The rate of heat conducted through a material depends on three parameters: the temperature difference ΔT between the two sides (°F), the surface area A of the material (ft²), and the material's R-value (ft² °F hr/Btu). The rate of heat transfer is given by the relationship:

$$\text{Heat} = \frac{A \Delta T}{R}$$

The R-values of materials typically vary throughout the structure, with different values for the floor, walls, and ceiling. Therefore, heat transfer must be calculated for each component and added together:

$$\text{Heat}_{TOTAL} = \text{Heat}_{FLOOR} + \text{Heat}_{WALLS} + \text{Heat}_{CEILING}$$

A unit of measure commonly used in the United States for the rate of heat transfer is British thermal units per hour (Btu/hr). For comparison, a 100-watt lightbulb produces 341 Btu/hr, which is about the average metabolic heat output of an adult person.

The following examples demonstrate a method for calculating conductive heat transfer in buildings.

EXAMPLE 2.1

A 600-square-foot home in northern Minnesota has 8-foot-high walls and a flat ceiling and rests on a concrete slab having a footprint of 20 feet by 30 feet. Rigid foam under the slab provides an insulation value of R-10. Insulation in the walls and ceiling establishes R-20 and R-40, respectively. Assume the home is maintained at a temperature of 70°F when the outside air temperature is constant at 0°F and the ground temperature beneath the slab is constant at 35°F. Determine the rate of conductive heat loss from the building. For simplicity, neglect the presence of windows and doors.

Solution:

STEP 1: Find the areas of the floor, walls, and ceiling.

$$A_{FLOOR} = \text{Length} \times \text{Width}$$
$$= 20 \text{ ft} \times 30 \text{ ft} = 600 \text{ ft}^2$$
$$A_{WALLS} = \text{Height} \times \text{Length}$$

where the total length is the perimeter of the building. If the building measures 20 feet by 30 feet, the perimeter equals 20 feet + 30 feet + 20 feet + 30 feet = 100 feet. The height is 8 feet, so

$$A_{WALLS} = 8 \text{ ft} \times 100 \text{ ft} = 800 \text{ ft}^2$$
$$A_{CEILING} = \text{Length} \times \text{Width}$$
$$= 20 \text{ ft} \times 30 \text{ ft} = 600 \text{ ft}^2$$

(continued)

(continued)

STEP 2: Compute the rate of heat loss for each component using the thermal conduction relationship.

$$\text{Heat} = \frac{A \Delta T}{R}$$

$$\text{Heat}_{\text{FLOORS}} = \frac{600\,\text{ft}^2 \times 35°\text{F}}{10\,\text{ft}^2\,°\text{F Btu/hr}} = 2100\,\text{Btu/hr}$$

$$\text{Heat}_{\text{WALLS}} = \frac{800\,\text{ft}^2 \times 70°\text{F}}{20\,\text{ft}^2\,°\text{F Btu/hr}} = 2800\,\text{Btu/hr}$$

$$\text{Heat}_{\text{CEILING}} = \frac{600\,\text{ft}^2 \times 70°\text{F}}{40\,\text{ft}^2\,°\text{F Btu/hr}} = 1050\,\text{Btu/hr}$$

STEP 3: Determine the total rate of heat flow by adding up the heat flow from each component.

$$\text{Heat} = \text{Heat}_{\text{FLOOR}} + \text{Heat}_{\text{WALLS}} + \text{Heat}_{\text{CEILING}}$$
$$= 2100\,\text{Btu/hr} + 2800\,\text{Btu/hr} + 1050\,\text{Btu/hr} = 5950\,\text{Btu/hr}$$

In order to maintain the home at a constant temperature of 70°F, a heat source (such as a furnace) must supply 5950 Btu/hr. Alternatively, if 17 people occupied the home, each outputting 341 Btu/hr of body heat, the house would remain at a nearly constant 70°F temperature.

EXAMPLE 2.2

Suppose the building of the previous example includes six windows with a U-factor of 0.40 and a total area of 100 square feet. Assume the building has two doors with R-values of 8.0 and a total area of 40 square feet. Find the rate of heat loss under the same conditions provided above.

Solution:

STEP 1: The areas of the windows and doors diminish the total area of the walls.

$$A_{\text{WALLS}} = A_{\text{ORIGINAL}} - A_{\text{WINDOWS}} - A_{\text{DOORS}}$$
$$= 800\,\text{ft}^2 - 100\,\text{ft}^2 - 40\,\text{ft}^2 = 660\,\text{ft}^2$$

STEP 2: Compute the new rates of heat loss. A U-factor of 0.40 translates into an R-value of $1/0.40 = 2.5$.

$$\text{Heat}_{\text{WALLS}} = \frac{660\,\text{ft}^2 \times 70°\text{F}}{20\,\text{ft}^2\,°\text{F Btu/hr}} = 2310\,\text{Btu/hr}$$

$$\text{Heat}_{\text{WINDOWS}} = \frac{100\,\text{ft}^2 \times 70°\text{F}}{2.5\,\text{ft}^2\,°\text{F Btu/hr}} = 2800\,\text{Btu/hr}$$

$$\text{Heat}_{\text{DOORS}} = \frac{40\,\text{ft}^2 \times 70°\text{F}}{8\,\text{ft}^2\,°\text{F Btu/hr}} = 350\,\text{Btu/hr}$$

(continued)

STEP 3: Determine the total rate of heat loss.

$$\text{Heat} = \text{Heat}_{\text{FLOOR}} + \text{Heat}_{\text{WALLS}} + \text{Heat}_{\text{CEILING}} + \text{Heat}_{\text{WINDOWS}} + \text{Heat}_{\text{DOORS}}$$
$$= 2100 + 2310 + 1050 + 2800 + 350 = 8610 \text{ Btu/hr}$$

The single largest rate of heat loss occurs through the windows. A U-factor of 0.40 is a common insulation value for double-paned windows used in many homes and businesses across the United States. The addition of windows dramatically undermines the insulation value of buildings and results in considerably greater heat loss.

EXAMPLE 2.3

Suppose all of the windows in the previous example are located on the south side of the building. During a sunny winter day, the amount of solar energy arriving at the windows averages roughly 275 Btu/hr per square foot of area during the 4 hours of midday light. How will the radiative heat gain through the windows during the day compare with their conductive heat loss? Assume the outside temperature remains at 0°F and the windows use clear glass with low-E coatings that produces an SHGC of 0.65.

Solution:

STEP 1: The rate of conductive heat loss through the windows remains unchanged from the previous example: $\text{Heat}_{\text{WINDOWS}} = 2800$ Btu/hr. The total amount of heat lost by the windows during a 24-hour period is equal to this rate of heat loss multiplied by 24 hours.

$$\text{Heat lost} = \text{Heat} \times \text{Time}$$
$$= 2800 \text{ Btu/hr} \times 24 \text{ hrs} = 67,200 \text{ Btu}$$

STEP 2: Determine the amount of solar heating during the 4 hours of sunlight.

$$\text{Heat gain} = \text{Heat} \times \text{Time}$$
$$= (\text{Solar intensity} \times \text{Window area} \times \text{SHGC}) \times \text{Time}$$
$$= \left(275 \, \frac{\text{Btu}}{\text{hr ft}^2} \times 100 \, \text{ft}^2 \times 0.65 \right) \times 4 \, \text{hrs} = 71,500 \, \text{Btu}$$

The amount of solar heating during the day more than compensates for the heat lost by the windows over a 24-hour period. This "solar gain" is very important in managing a building's overall energy demands. Building configurations that allow winter sunlight to warm interior spaces are an important aspect of passive solar design, which is the subject of the next chapter.

Chapter Summary

Historically, building materials came from locally harvested and minimally processed resources. Advances in technology and the availability of fossil fuels allowed energy-intensive manufacturing techniques and long-distance transportation to create and distribute a vast array of building materials to sites across the globe. While these advancements have created opportunities for the built environment that were not previously possible, they have also resulted in an industry that consumes more resources and uses more energy than ever before. Green building makes an attempt to select materials that can be harvested sustainably and to minimize environmental degradation.

Although modern manufacturing processes are often energy intensive, they can lessen the consumption of natural resources through improved resource utilization and reduced waste. The benefits include:

- Processing low-grade, fast-growing trees into OSB, fiberboard, engineered siding and flooring, laminated beams and posts, and an array of other products, all of which reduce demand on remaining old-growth forests.
- Treating lumber with preservatives that extend the useful life of wood in exterior or wet conditions and allow fast-growing trees to replace slow-growing species like cedar and redwood.
- Creating advanced insulation products that allow homes to be tighter and more comfortable and to use less energy.
- Coating window glass with thin films that select for particular wavelengths, transmitting visible light but blocking the passage of heat.

- Implementing innovative recycling techniques that allow the reuse of metals, paper, glass, and plastics, many of which find their way into the building industry as recycled metal roofing, cellulose insulation, window glass, laminate flooring, PVC siding, and many other products.

These advancements, along with many others, have not been without their problems, however. Many of the engineered wood products contain adhesives that outgas VOCs. Toxins present in building materials and pollution caused during their manufacture contribute to many environmental problems and public health concerns. The production of electricity—essential for manufacturing—comes primarily from the burning of coal, which generates sulfur dioxide, nitrogen oxides, mercury, and carbon dioxide. These cause air and water pollution, acid rain, and smog and contribute to climate change.

In designing a green building, the benefits and problems of materials need to be carefully considered to minimize the impacts on the environment and promote healthy living spaces for people and the surrounding communities.

Review Questions

1. Materials such as metal, tile, and cement have large embodied energy due to mining and high-temperature processing. Nevertheless, these materials are often considered good choices for a green building. Why?

2. What are some of the advantages and disadvantages of using prefinished wood siding compared to wood siding that must be finished on site?

3. Consider the benefits and drawbacks of using plywood compared to oriented strand board for building sheathing. Which product would you consider the better choice? Explain your reasoning using principles of green building.

4. Suppose a builder wishes to avoid the use of concrete during the construction of a house. Instead of a concrete foundation, the builder selects a permanent wood foundation made

of pressure-treated lumber laid on top of well-drained, locally sourced gravel. Use the principles of green building to assess the merits and drawbacks of this form of foundation.

5. The designs of many cold climate buildings include an entry room that is closed off from the main living spaces by a door. Explain how an entry room may improve the performance of a building.

6. Explain the difference between vinyl and linoleum flooring. What are the advantages and disadvantages of each?

7. Rank the embodied energy for asphalt shingles, steel nails, and cellulose insulation. Consider the extraction of the raw material, processing, transportation, and other factors. Explain your reasoning.

8. Shortly after moving into a newly constructed house, a homeowner reports frequent

headaches, dizziness, and nausea, none of which were experienced previously. What might be the cause of these symptoms, and what practical remedy would you suggest?

9. A homeowner residing in a warm, wet climate wishes to build an exterior deck for a family that includes three young children, one of which is barely crawling. What recommendation would you make for the decking material? Your consideration may include materials not mentioned in this text. Explain your reasoning.

10. Suppose you purchase an old, drafty home with hardwood floors, single-paned windows, vinyl siding, and asphalt roofing. What measures would you initially take to make the home as green as possible? Assume a modest budget, and explain your reasoning.

11. Describe the three forms of heat transfer that occur around a single lit candle.

12. How does R-value relate to the three forms of heat transfer?

13. If 3000 Btu/hr of heat pass through a wall that has an insulation value of R-10, how much heat will pass through an identical wall with an insulation value of R-20?

14. A homeowner insulates the attic of a cold climate home with R-38 fiberglass batts. During the summer months, the fiberglass performs to its rated insulation value. However, during the winter months, the fiberglass proves to be less effective. Why might this be? What could the homeowner do to remedy this? What may have been a better choice of insulation?

15. Explain how each of the four window parameters (U-value, SHGC, VT, and air infiltration) relates to the three forms of heat transfer (conduction, convection, and radiation).

16. What factors determine the number, type, and placement of windows for a green building?

17. Which insulation types have relatively large recycled content? What are the disadvantages of these insulation types? What are the advantages?

18. What advantages do asphalt shingles have over clay tiles? What disadvantages?

19. If you were designing a green house in a fire-prone area in California, what type of siding, roofing, and flooring would you choose? Explain why your choices uphold the principles of green building.

20. In what climates are windows with low-E coatings appropriate? In what climates are windows with low SHGC appropriate?

Practice Problems

1. What is the R-value of a double-paned window that has a U-factor of 0.33? What percentage of heat loss will occur at night if a 2-inch sheet of rigid EPS insulation tightly covers the window?

2. A properly installed window quilt can provide an additional insulation of R-4 to windows. For a window with a U-factor of 0.33, determine the percentage of heat lost for a quilt-covered window compared to a bare window.

3. A cubic ice chest measuring 2 feet on a side contains a block of ice at 32°F. If the ice chest is made of 2-inch-thick EPS foam and the outside temperature is 74°F, what is the rate of heat conduction into the ice (in Btu/hr)?

4. A chicken coop is heated with an incandescent lightbulb that provides up to 500 Btu/hr of heat. An automatic thermostat set at 60°F controls the output of the lightbulb. The coop is closed at night and is uniformly insulated to R-10. If the nighttime temperature is 20°F, what is the maximum surface area of the chicken coop (in ft^2) in order for the lightbulb to maintain a constant internal temperature of 60°F?

5. A ballet class of 10 dancers uses a discarded shipping container as a temporary classroom. The walls, floor, and roof of the container are made of the same uniform material, and the container measures 10 feet × 10 feet × 30 feet. On a day when the air temperature is 30°F and the ground temperature is 40°F, determine the minimum R-value of the container in order to maintain a constant interior temperature of 65°F. Assume each of the 10 dancers outputs 600 Btu/hr of metabolic heat and no additional heat source is available.

Passive
Solar Design

Courtesy of Scott Grinnell.

Introduction

The daily and seasonal motions of the sun have influenced building design since the earliest civilizations. Throughout history people have oriented and designed their buildings to make use of the sun's energy and light. Passive solar design utilizes seasonal changes in the path of the sun to create structures that are heated (and sometimes cooled) by natural means.

As early as 400 B.C., the ancient Greeks used passive solar design in their buildings and planned entire cities to ensure every citizen had access to sunlight. The Romans later improved on these designs, creating glass for windows, inventing greenhouses, and passing laws forbidding structures from shading a neighbor's residence. Native Americans, such as the Anasazi, built dwellings beneath cliff overhangs that offered shade during the summer but welcomed the winter sunlight (Figure 1.1). Early New Englanders favored a saltbox design with two stories of south-facing windows and a north roof that sloped down to a single story with few windows (Figure 3.1). The Spaniards who settled in the American Southwest oriented their homes east-to-west so that the long axis faced south and used roof overhangs and decks on the second floor to shade windows from the summer sunlight (Figure 3.2). These features all represent aspects of passive solar design.

The Industrial Revolution and the advent of inexpensive fossil fuels changed building design, allowing people to heat or cool nearly any building without relying on the influence of the sun. Site-specific designs gave way to standardized structures that spread worldwide across vastly different climates. Today it is common to find homes designed with no regard whatsoever for the sun's path. These homes, which use more fossil fuels than those incorporating passive solar design, tend to be energy and resource inefficient and create spaces that are less comfortable for the inhabitants. Interior spaces are often too hot or too cold, suffer from excessive glare, are too dim or too gloomy, or lack a sense of connection to the outside.

FIGURE 3.1 The saltbox design, with two stories of south-facing windows and a long sloping roof protecting the north side, was common in New England during the 1700s. The cold winter climate made heating a priority. *Source:* Library of Congress, Prints & Photographs Division, HABS MASS, 11-QUI, 7-1.

FIGURE 3.2 The Spanish colonial design, with generous overhangs protecting living spaces from sunlight, was common in the American Southwest. The desert climate made summer cooling a primary concern. *Source:* Library of Congress, Prints & Photographs Division, HABS NM, 24-LAVEG.V, 1-2.

Because passive solar design utilizes seasonal changes in the path of the sun to harness natural renewable energy, understanding the daily and seasonal motions of the sun is essential.

Seasonal and Daily Paths of the Sun

The earth rotates once each day about an axis that is tilted 23.5° with respect to its orbit around the sun. As shown in Figure 3.3, the direction of tilt remains fixed as the earth moves through space, changing the earth's orientation with respect to the sun. This is what creates the seasons. When a hemisphere tilts toward the sun, it experiences longer, warmer days; when it tilts away, it experiences shorter, cooler days. On the spring and fall **equinoxes** (around March 20 and September 22), the earth's tilt is in the same direction as its orbital motion, so neither hemisphere tilts toward or away from the sun. As a result, every place on the earth experiences the same length of day. Around June 21, the

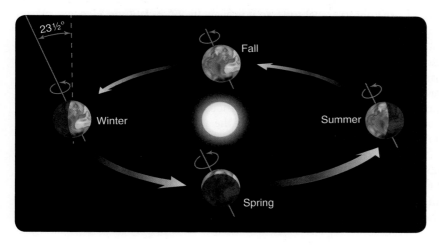

FIGURE 3.3 Although the tilt of the earth's axis remains constant during its yearly orbit around the sun, the orientation of its tilt with respect to the sun changes, producing the seasons. The hemisphere tilted toward the sun experiences summer, while the hemisphere tilted away experiences winter. Spring and fall occur when the tilt is parallel to the orbital direction. © 2016 Cengage Learning®.

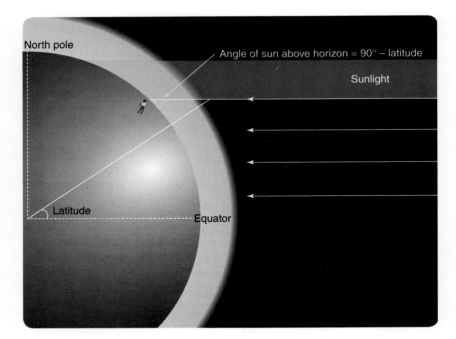

FIGURE 3.4 During the spring and fall equinoxes, the sun's rays strike the earth perpendicular to the equator. An observer at the equator will see the midday sun directly overhead, a solar altitude of 90°. For observers at some latitude north or south of the equator, the solar altitude is given by 90° − latitude.

© 2016 Cengage Learning®.

Northern Hemisphere tilts directly toward the sun and experiences its summer solstice, while the Southern Hemisphere experiences its winter solstice. On about December 21, the reverse is true: the Northern Hemisphere tilts directly away from the sun, producing its winter solstice and the Southern Hemisphere's summer solstice.

Observers on earth perceive these seasonal variations as differences in the maximum height of the sun above the horizon, called the altitude. The sun's altitude varies not only with season but also with latitude. Consider Figure 3.4, which depicts the earth during the spring or fall equinox. On the equinoxes, the sun's rays are perpendicular to the equator, and someone residing on the equator would see the midday sun directly overhead. For locations north or south of the equator, the sun's altitude is lower in the sky, reduced by the observer's latitude, so it is given by 90° − latitude. Hence, a person residing at 40° N latitude would see the noontime sun 50° (90° − 40°) above the horizon. On the other hand, a person at the North or South Pole would see the sun's rays right on the horizon (at 0°, or 90° − 90°).

During the summer solstice, the sun is 23.5° higher in the sky than during the equinoxes, and the sun's altitude is given by 90° + 23.5° − latitude (Figure 3.5). A person residing at 40° N latitude would now observe the noontime sun at 73.5° (90° + 23.5° − 40°) above the horizon. During the winter solstice, the angle of the sun is 23.5° lower in the sky, or 90° − 23.5° − latitude, dropping the noontime sun for a 40° N latitude observer to 26.5° (90° − 23.5° − 40°) above the horizon. Hence, over the course of a year the sun's altitude varies by 47°. This annual variation follows a sinusoidal curve, shown in Figure 3.6, changing daily in a subtle, predictable manner. This predictability has allowed people throughout the ages to build structures that maximized the benefit of the sun during the course of the year.

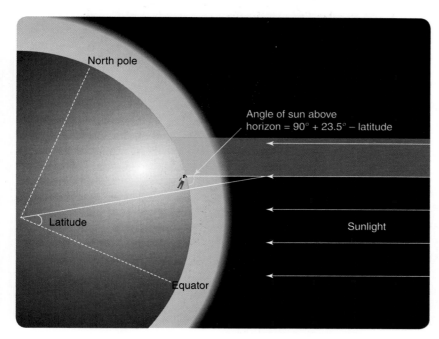

FIGURE 3.5 Around June 21, the Northern Hemisphere is tilted 23.5° toward the sun. Now Northern Hemisphere observers see the sun at a solar altitude of 90° + 23.5° − latitude. Residents in the Southern Hemisphere experience winter, and their observed solar angle is reduced by 23.5°. © 2016 Cengage Learning®.

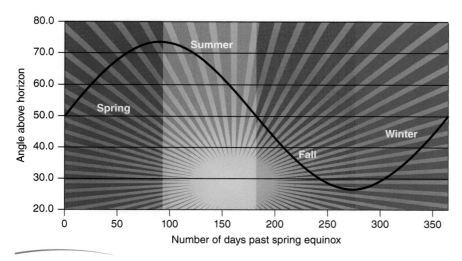

FIGURE 3.6 The solar altitude for any particular location (shown here at 40° N latitude) varies predictably over the course of the year, following a sinusoidal curve. On the spring equinox, the solar altitude is 90° − 40° = 50°. Ninety-one days later, during the summer solstice, the solar altitude reaches its maximum of 90° + 23.5° − 40° = 73.5°. The solar altitude returns to 50° on the fall equinox and then decreases to a minimum of 90° − 23.5° − 40° = 26.5° on the winter solstice. Although the observed altitude depends on latitude, every location experiences a 47°-variation over the course of the year. © 2016 Cengage Learning®.

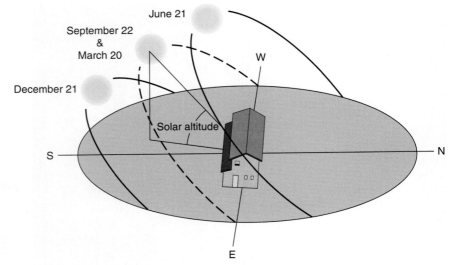

FIGURE 3.7 Seasonal variations in the sun's altitude also change the locations of the rising and setting sun. Notice that light from the winter sun is predominantly from the south, while the summer sun illuminates all sides of a building. © 2016 Cengage Learning®.

Designing a building to benefit from the seasonally changing altitude of the sun is a fundamental element of passive solar design. Understanding this change allows a builder to create comfortable and energy-efficient spaces.

Figure 3.7 shows the paths of the sun for the summer solstice, winter solstice, and spring and fall equinoxes for a Northern Hemisphere home. The winter sun is relatively low in the sky, and its light comes almost entirely from the south. During the winter, very little direct sunlight penetrates windows on the east and west sides of buildings, and none comes through north windows. The high summer sun, on the other hand, rises and sets north of

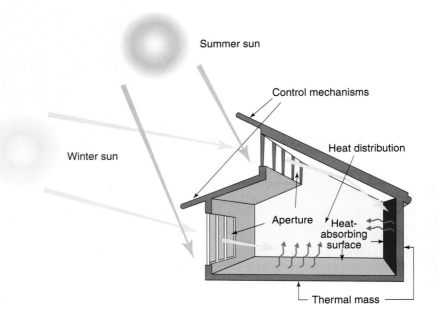

FIGURE 3.8 Passive solar design allows buildings to harness solar energy from the low winter sun, while preventing heat gain from the high summer sun. Passive solar design consists of five key elements: aperture, heat-absorbing surface, thermal mass, heat distribution, and control mechanisms.
© 2016 Cengage Learning®.

the east-west line, allowing significant illumination through east and west windows and even some direct light through north windows. At all times of the year, the sun's altitude remains relatively stable as it approaches and descends from its maximum height at noon each day. As a result, roof eaves and awnings are able to shade south-facing windows from the rays of a relatively high summer sun but allow those of the lower winter sun to pass through. In this way, heat enters the building only when needed, as shown in Figure 3.8.

Passive Solar Design

Passive solar design is best known for harnessing the sun's warmth during the winter months, while preventing unwanted heat gain in the summer. However, to some extent passive solar design is also able to cool buildings directly during the summer by establishing sun-induced ventilation. Passive solar design can, therefore, reduce energy use (and its associated environmental impacts) for both heating and cooling. In addition, passive solar design saves money by allowing smaller heating, ventilation, and air conditioning units, provides insurance against spikes in conventional energy costs, offers greater comfort and connectedness to the outdoors, and reduces lighting needs and costs.

Although its principles can be applied universally, passive solar design is site specific and must be adapted to the local climate. Cold and temperate climates, where winter heating is the dominant concern, require strategies that promote generous solar heat gain, typically accomplished through south-facing windows. In regions where cooling is the dominant concern, appropriate strategies include preferential use of north-facing windows, ample ventilation, and comprehensive shading from both direct and reflected sunlight.

Successful passive solar buildings must be well insulated, oriented to allow solar access, and able to adequately collect, control, store, and distribute solar energy. A building design that is entirely passive relies only on conduction, convection, and radiation to distribute heat throughout the interior of a building. Many buildings that employ passive solar design elements also include ducts, temperature-controlled blowers, and other mechanical means to distribute heat. The addition of these systems, while often beneficial, creates buildings that are no longer strictly passive solar.

Passive solar design relies on the five key elements listed in Table 3.1 and shown in Figure 3.8. Each of these elements is essential. The absence of any element may render the design ineffective and eliminate the benefits of the entire system.

If not properly managed, sunlight entering a building can make interior spaces unbearably hot even on a frigid winter day. In fact, a darkly painted box, insulated and sealed behind glass, can reach an interior temperature of 300°F or higher on a cold, sunny winter day. As soon as the sun disappears, however, the temperature in the box rapidly cools. To a lesser extent, the same large temperature fluctuations will occur inside a solar-heated building that lacks appropriate heat-absorbing surfaces and thermal mass.

Aperture

Windows are the most common form of aperture. The appropriate size, number, and type of windows depend on the climate and intended use of the building. For climates that require heating, windows should preferentially be located on the sunward (south) side and have large solar heat gain coefficients (SHGCs) to permit plentiful heat gain

TABLE 3.1 Passive solar design elements.

ELEMENT	EXAMPLES
Aperture	• Window • Cavity opening
Heat-absorbing surface	• Wall or floor • Black metal baffle • Barrel • Dark pipes or hoses
Thermal mass	• Concrete • Stone • Brick • Sand • Water
Heat distribution	• Arrangement and coloration of interior partitions • Surface conduction and radiation • Vents or ducts • Interior windows and doors
Control mechanisms	• Roof eaves and awnings • Landscaping • Window coatings, coverings, and sunscreens • Pergolas

from the winter sun. Windows on the east, west, and north sides, however, should have low SHGCs because they are primarily illuminated by the summer sun when heat gain is usually undesirable. For hot climates where cooling is the dominant concern, all windows should have low SHGC ratings and may additionally be tinted. In any climate, low-E (low-emissivity) coatings on windows are beneficial whenever buildings themselves are conditioned, either by heating or by cooling. Low-E coatings limit the passage of heat by radiation, thereby blocking interior heat from leaking out into a cold, wintry exterior or blocking exterior heat from leaking into a cool interior on a hot day.

Heat-Absorbing Surface

The heat-absorbing surface is the surface that the sunlight strikes after it has passed through the aperture. Dark-colored surfaces generally perform better by preventing sunlight from reflecting back out through the aperture. This surface may be the glazed tile on the floor of a building, the paint on a wall, the side of a barrel (that may contain water), or any other object capable of absorbing sunlight.

Thermal Mass

Once the heat is absorbed, it needs to be retained. Thermal mass allows the radiant energy absorbed during the day to be stored and released gradually during the night, making the temperature much more stable and the building more comfortable. Thermal mass is any massive material that can store large amounts of thermal energy, such as a concrete slab, stone or tile flooring, masonry walls, or water stored in a barrel. The amount of thermal mass required in a particular situation depends on the amount of solar energy absorbed during the day and the length of time the material must supply heat.

Heat Distribution

An effective solar design requires a means to distribute heat comfortably throughout a building. This can be accomplished by radiation from warmed surfaces; by convective air movements through windows, doors, or vents; or directly by conduction across materials. Buildings with open floor plans can often rely solely on passive means (natural conduction, convection, and radiation), which results in a true passive solar design. Buildings with numerous isolated rooms, on the other hand, often require mechanical systems, such as fans or pumps, to help distribute heat.

Control Mechanism

Control mechanisms regulate when and how much solar energy enters a building. The most common control mechanisms in passive solar design are awnings and roof eaves. The size and location of windows and the geometry of awnings and eaves determine when sunlight enters the building. As shown in Figure 3.9, the amount of sunlight entering a space changes over the course of the year. The use of roof eaves as a control mechanism entails a compromise due to the symmetry of the sun's altitude around the solstice: eaves that shade a window in late summer, when heat is generally unwelcomed, will also do so in spring, when solar heating is usually desirable. Typically, cold climate designs should allow for spring heating and use other control mechanisms, such as deciduous landscaping, to shade south-facing windows in the late summer. Alternatively, the use of adjustable or removable awnings to shade windows eliminates this compromise altogether. Awnings that allow access of all winter and spring sunlight can be repositioned to block summer and early autumn sunlight.

Deciduous landscaping is also a beneficial control mechanism for the east and west sides of buildings. The low rays of the summer sun in the morning and afternoon are difficult to shade, making overhangs ineffective for east- and west-facing windows. In the late spring, when conditions are still cool, trees often have not fully leafed out and allow sunlight to enter east and west windows. Conversely, in the summer and early fall, when conditions remain warm, these same trees hold their leaves and block sunlight from entering the building, preventing solar heat gain.

The north side of buildings (in the Northern Hemisphere) receives very little sunlight and requires no particular control mechanism for solar energy. However, landscaping on the north side can block cold winter winds, encourage ventilation and cooling air movement, and enhance the comfort of occupants in other ways, depending on climate. Therefore, appropriate landscaping is an important consideration for all sides of a building.

As discussed in Chapter 2, windows are relatively poorly insulated and vulnerable to unwanted heat transfer. Because of this, passive solar design often calls for additional control mechanisms to prevent unwanted heat transfer through windows. These include a variety of window coverings, which depend on climate and need. In general, cold climates require window coverings on the interior of the window, while hot climates require them on the exterior.

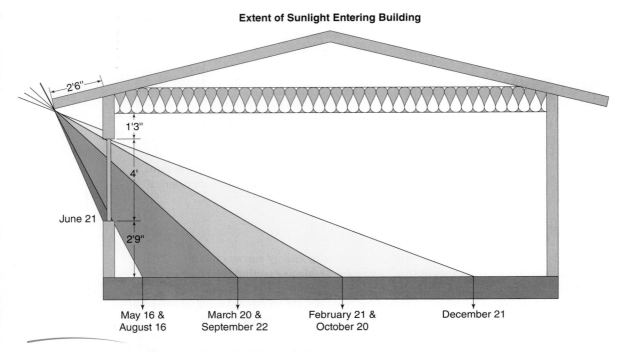

Extent of Sunlight Entering Building

FIGURE 3.9 The length of the eaves and the size and placement of the windows determine the dates when sunlight enters south-facing windows. This example shows a passive solar design for a home in northern Wisconsin (46.5° N latitude), discussed further in Case Study 3.1. No direct sunlight enters the building between May 16 and August 16. By October 20, the windows are fully illuminated and remain so until February 21, when the eaves gradually eclipse some of the sunlight. © 2016 Cengage Learning®.

In climates where winter heating is required, abundant use of south-facing windows relies on the principle that the amount of heat gained during a sunny winter day exceeds the amount of energy lost through the windows over a period of 24 hours or longer. However, for regions that are often cloudy, this may not always be the case without further insulating windows at night. Window quilts, cellular shades, and other tight-fitting insulated interior window treatments minimize convection, conduction, and radiation. These window treatments can be rolled up during the day and lowered at night, either manually or with remotely operated motors, and can cut heat loss through windows by a factor of three or more. This can result in a 20%–30% increase in annual solar performance for cold climate buildings. Window quilts and cellular shades come in a variety of styles, textures, and colors and can be adapted to most windows.

Buildings in climates with hot summers can benefit from exterior shutters and shade screens that minimize heat gain during the day. Shutters not only block out light but also provide security and protection from severe weather, including hurricanes. They can be operated manually or with electric motors, opening and closing in overlapping plates much like modern garage doors. However, they block out all light and may be undesirable for many applications. Shade screens, which can replace external window screens, can significantly reduce glare, ultraviolet light, and solar heat gain, while imposing only modest reductions in visibility. Shade screen material permits air circulation and functions as regular screening for stopping insects.

Types of Passive Solar Systems

Three types of passive solar systems facilitate a variety of design options, each offering different benefits and limitations (see Table 3.2). A single building may utilize several different passive solar systems for different spaces. The appropriate system depends on

TABLE 3.2   Advantages and disadvantages of passive solar systems.

Direct Gain (Heating Only)

Advantages	• Comparatively inexpensive • Provides light to interior spaces • Offers visual connection to outdoors • Operable windows provide ventilation
Disadvantages	• Windows are relatively poorly insulated • Maximum heating occurs during the day • Possible UV damage to interior spaces • May result in unwanted glare

Indirect Gain—Unvented Trombe Wall (Heating Only)

Advantages	• Maximum heating occurs at night when most needed • Moderates daily temperature swings • Useful for spaces where direct light is undesirable
Disadvantages	• Lacks visual connection to outdoors • Low insulation value limits applicability to climates with regular daily sunshine and moderate temperatures

Indirect Gain—Thermosiphon Including Vented Trombe Wall (Heating or Cooling)

Advantages	• Can provide both heating and cooling • Cooling can occur during the day when most needed • Offers natural ventilation • Can utilize evaporative cooling
Disadvantages	• Vents must be closed to prevent unwanted back drafting • Cooling mode requires cool air source (e.g., ground), which may not be available when most needed • Evaporation requires source of water

Sunspace (Heating or Cooling)

Advantages	• Can serve as a greenhouse or other auxiliary space • Isolated from building • Can be added after building is complete • Can be designed for nearly any climate
Disadvantages	• Living space receives only indirect light • Vents, doors, or windows must be opened and closed at appropriate times for optimal heat transfer

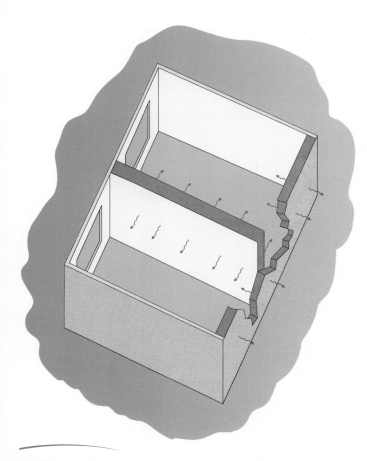

FIGURE 3.10 Thermal mass located within the building envelope performs better than thermal mass in an exterior wall. Both sides of an interior wall radiate to the inside compared to only one side of an exterior wall. © 2016 Cengage Learning®.

the intended use of the space, whether the space requires heating or cooling, and the local climate. The three types of systems are direct gain (for heating only), indirect gain (for heating or cooling), and sunspaces (for heating or cooling).

Direct Gain

Direct gain systems are the most common and simplest type of passive solar design. Intended to provide heat, direct gain systems use south-facing windows, appropriate controls, and adequate thermal mass to directly warm interior spaces with sunlight during the day. They tend to be the least expensive type of system to incorporate into building design and work well in spaces that can tolerate bright illumination. Although thermal mass moderates temperature swings, direct gain systems provide the bulk of the heat when the sun is shining, which is not always the time when heat is most needed. In addition, sunlight may fade materials and create undesired glare.

Direct gain systems require sufficient thermal mass to absorb and store solar energy. The capacity of thermal mass to store thermal energy increases with thickness up to about 5 inches, beyond which additional benefits are typically small. The absorbing area should be five times or more the aperture area, though not all of the thermal mass must be directly illuminated by sunlight, as heat naturally transfers through the material. Establishing sufficient thermal mass is a design element that must be determined early in the building process.

When direct gain systems utilize floors as the primary mass storage, the floor surface should be dark, while walls and ceilings should be light. This allows the floor to absorb and radiate energy and the walls and ceilings to reflect and distribute warmth and light throughout the building. When walls serve as the primary thermal mass, they should be dark. Massive internal partitions function better than massive external walls because both sides of the internal partition can participate in heat storage and radiate into the building, as shown in Figure 3.10. Regardless of location, all thermal mass should be left uncovered, free of rugs and furniture and wall hangings.

Indirect Gain

Indirect gain systems heat a surface exterior to the living space and use natural forms of heat transfer to transport heat to the interior spaces. Indirect gain systems include thermal storage walls and thermosiphons.

Thermal Storage Walls

A **thermal storage wall**, or **Trombe wall**, collects and stores solar energy during the day and releases it gradually to the interior during the night (Figure 3.11). The thickness and conductivity of the Trombe wall are critical, as the wall must be massive enough to absorb available solar energy and sufficiently conductive to begin delivering heat shortly after sunset.

Trombe walls are massive, south-facing walls typically composed of concrete, stone, adobe, or other masonry and insulated from the outdoors by a single or double glazing of glass. An air gap of an inch or more separates the wall from the window glazing and allows visible sunlight to warm the wall. Selective coatings on the windows and Trombe walls can improve solar absorption and limit heat loss at night and during cloudy days. The wall prevents sunlight from directly illuminating the living space and can be used for places that need to be kept dark or where views to the south are undesirable.

Trombe walls themselves are uninsulated and, therefore, are effective only in climates with regular daily sunshine, such as deserts, that may experience large *daily* temperature swings but undergo relatively moderate *seasonal* change. Buildings often benefit from a combination of systems: for example, the immediate warmth of natural windows for one space with the delayed warmth of a Trombe wall for another. However, in cold climates or those without consistent sunshine, direct gain systems with well-insulated walls are much more practical.

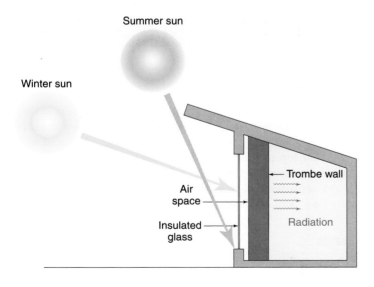

FIGURE 3.11 The thermal mass of a Trombe wall warms gradually during the day and radiates heat by nightfall, providing warmth during the night and moderating daily temperature swings. © 2016 Cengage Learning®.

Further Learning TRADITIONAL ADOBE HOMES

Traditional adobe homes built in the American Southwest operate in much the same manner as Trombe walls by absorbing heat during the day and releasing it gradually during the night. Solid adobe walls offer rather poor insulation but provide excellent thermal mass, effectively moderating daily temperature swings by delaying heat from entering the building until after sunset. This principle guided the Pueblo Indians in designing their plateau-top city of Acoma, one of the oldest continuously inhabited communities in the United States (see Figure 3.12). Warmed by the winter sun during the day, the adobe walls deliver heat after nightfall. The higher summer sun less effectively warms the south walls because the sun's rays strike the walls at a more oblique angle. The rooftop, which receives the brunt of the summer sun, is insulated to limit heat gain.

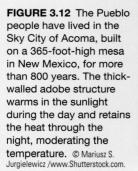

FIGURE 3.12 The Pueblo people have lived in the Sky City of Acoma, built on a 365-foot-high mesa in New Mexico, for more than 800 years. The thick-walled adobe structure warms in the sunlight during the day and retains the heat through the night, moderating the temperature. © Mariusz S. Jurgielewicz /www.Shutterstock.com.

As with direct gain systems, the best method for reducing unwanted heat gain with Trombe walls is a design including awnings or appropriate overhangs to create shade in the summer, when heating is undesirable. Additionally, some walls can be fitted with seasonally adjustable vents on the top and bottom of the glazing to better allow summer heat to escape to the outside, as shown in Figure 3.13. However, since Trombe walls are uninsulated, climates with consistently high summer temperatures generally require additional measures to limit heat gain—such as exterior insulated shutters.

Thermosiphons

Another form of indirect gain is a **thermosiphon**, a natural convective system that relies on sunlight to heat and move air without mechanical assistance. Thermosiphons can either warm or cool a building, depending on the system. For either type of system, solar

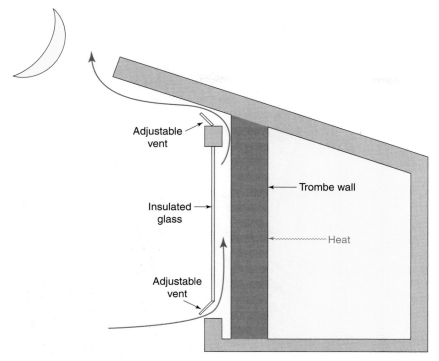

Adjustable vent

Trombe wall

Insulated glass

Heat

Adjustable vent

FIGURE 3.13 An exterior-vented Trombe wall creates a natural circulation at night. The warm wall heats adjacent air, which rises and exits through the top vent. Cool air flows in the lower vent, creating a circulation that draws heat away from the wall and cools the interior. © 2016 Cengage Learning®.

energy heats an absorbing surface that warms the surrounding air. The heated air rises, drawing in cool air to replace it and thus creating a convective loop.

A closed-loop thermosiphon system warms a building by drawing cool air from the building, heating it, and delivering the warmed air back to the building (Figure 3.14). This type of system offers a method of warming a space during the day without exposing the interior to direct sunshine.

An open-loop thermosiphon system offers ventilation and may cool a building by drawing in cooler air from the exterior and allowing heated air to escape, as in a solar chimney (Figure 3.15). A solar chimney can be as simple as an open absorber on the roof that connects via a chimney to air inside the building. As heated air rises from the absorber, it draws air into the building to replenish the air escaping through the chimney. However, in order for a solar chimney to cool a building, rather than simply providing ventilation, it must draw in air that is cooler than the interior air. This often limits the effectiveness of solar chimneys. Air must be drawn from shady, cooler places or through tubes buried underground where temperatures are lower. However, these systems require careful design to prevent entry of insects, mice, and other vermin and are limited to regions that are free of radon and other unhealthy soil gases. An additional limitation is that all thermosiphon systems will reverse direction at night unless one-way dampers prevent back drafting.

Solar chimneys can be particularly effective in dry climates by moistening the incoming air. The evaporation of water cools the air, often substantially, but requires an ample supply of water, which is not always available.

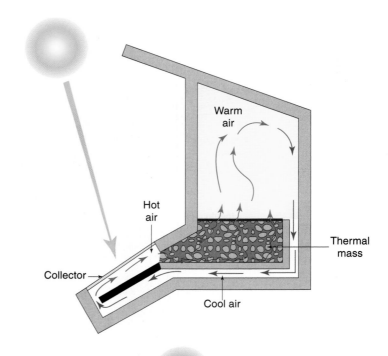

FIGURE 3.14 Thermosiphon systems use collectors exterior to the building and rely on ducts and vents to distribute heat to the interior. In this closed-loop arrangement, sunlight heats air that circulates through the building, delivering warmth during the day. © 2016 Cengage Learning®.

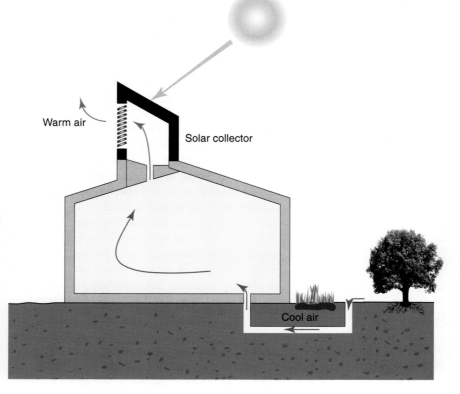

FIGURE 3.15 Thermosiphon systems can cool buildings during the day by heating air in a solar chimney. Hot air within the chimney rises and draws warm air from the building interior. Cool fresh air flows in to replace the exhausted air. In dry climates, adding moisture to the incoming air further cools the building through evaporation. © 2016 Cengage Learning®.

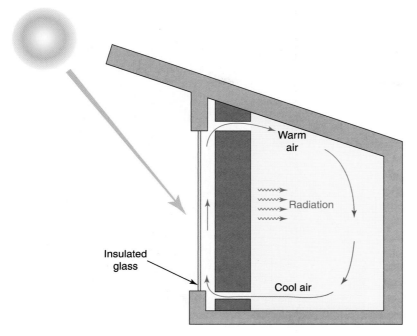

FIGURE 3.16 A Trombe wall vented to the interior acts like a closed-loop thermosiphon system. The warmed air rises through the top vent and draws cooler air through the lower vent, heating the interior during the day rather than the night.
© 2016 Cengage Learning®.

Trombe walls can be modified to behave as a thermosiphon system by adding vents at the top and bottom of the wall, as shown in Figure 3.16. These vents allow interior air to circulate behind the wall and be heated by the sunlight. The warm air rises and enters the building interior through the top vent, and cool air enters through the lower vent to replace the heated air. This establishes a natural convective loop without mechanical assistance. However, adding vents to Trombe walls delivers heat to the interior while the sun is shining, rather than after the sun has set, and limits thermal energy stored for nighttime heating. Furthermore, vents must be closed at night, or the Trombe wall will work in reverse: cooling a warm interior. Vents also permit the entry of dust and other contaminants that tend to degrade the performance of Trombe walls. Nevertheless, vented Trombe walls can be useful by providing daytime heating without the glare associated with windows.

Sunspaces

The third type of passive solar system is sunspaces, such as solariums, conservatories, or attached greenhouses, which combine direct gain and indirect gain systems. Sunspaces are sometimes called isolated systems because they are able to function independently of the building interior.

In this type of system, the sun directly heats a sunspace through abundant south-facing windows. The sunspace then shares some of its warmth with the adjoining building either by conduction through a massive wall or by convection through openings, as shown in Figure 3.17 and Figure 3.18. Thermal mass—in the floor, the shared wall, or even barrels of water—moderates temperature swings. Nevertheless,

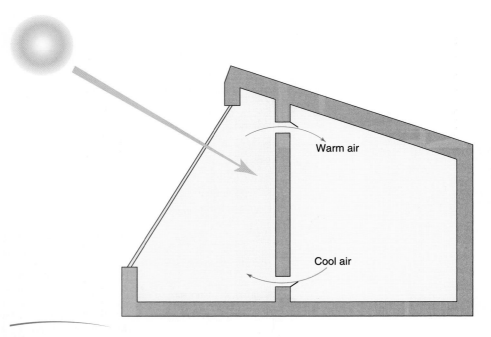

FIGURE 3.17 During the day, sunspaces heat an adjoining building by convection through appropriately located vents (as shown) or by conduction through a Trombe wall. At night, sunspaces can become quite cold and can be used to cool buildings. Whether sunspaces heat or cool adjoining spaces depends on venting. © 2016 Cengage Learning®.

FIGURE 3.18 This attached greenhouse shares its warmth with an adjoining farmhouse in northern Wisconsin. Courtesy of Clare Hintz, Elsewhere Farm.

sunspaces often experience extreme temperatures due to significant solar heat gain during the day, followed by significant heat loss through windows at night. This can compromise energy efficiency and comfort within the adjoining building unless the two spaces can be adequately separated. Closing interconnecting windows and doors may be all that is required to prevent unwanted heat gain during the day or heat loss during the night. In mild, sunny climates with large daily temperature swings, the wall between the sunspace and the adjoining building can be designed like a Trombe wall—massive and uninsulated—to allow conduction to transfer heat gradually throughout the night. This helps prevent overheating during the day and overcooling at night but is less effective in cold or cloudy climates where sunshine is inadequate to compensate for heat lost through the uninsulated wall.

Active Systems

Although passive solar design relies on the transfer of heat without mechanical assistance, many builders choose to augment passive solar design with active solar systems. These systems can be very effective in delivering heat to spaces more remotely located in the building. For example, vented Trombe walls and thermosiphon systems may include blower fans to distribute air or pumps to circulate water. When powered by solar electricity (discussed in Chapter 7), these systems can become energy-neutral, adding comfort without increasing the electrical demands of the building. Ceiling fans, which augment natural heat distribution, are also common additions to passive solar structures.

CASE STUDY 3.1

Passive Solar Home

Location: Northern Wisconsin (46.5° N latitude)

Year built: 2011

Size: 1750 ft^2

Climate: The average monthly temperatures vary from cold in the winter to comfortable in the summer. The amount of sunshine varies as well, with moderate cloud cover all year long. During the winter, particularly the months of November and December, cloud cover can linger for weeks at a time amid subfreezing temperatures. Heating during the winter is the primary concern.

System Design Preference: Thermal storage walls would be ineffective in this climate because of the extreme cold and extended periods of cloud cover. Thermosiphon systems for cooling are unnecessary due to the relatively mild summers. The owners selected a direct gain passive solar design in a well-insulated structure.

Aperture: The aperture consists of 11 large windows on the south side of the house, all of which are awning style and triple-paned to provide an insulation value of U = 0.24. While all

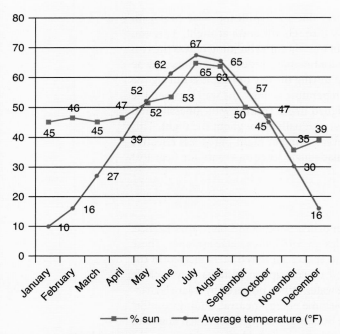

The average monthly temperature and percentage of sunshine indicate that buildings in northern Wisconsin require supplemental heat most of the year. During the particularly cloudy months of November and December, a passive solar design is insufficient to maintain a comfortable temperature. © 2016 Cengage Learning®.

Taken on the winter solstice, this photo reveals shadows from nearby deciduous trees. As the trees grow, they may limit passive solar heating. Courtesy of Scott Grinnell.

windows include low-E coatings, windows on the south side are otherwise untreated and offer an SHGC of 0.57 (allowing 57% of the solar energy into the house). Windows sparingly placed on the north, east, and west sides of the building include additional coatings that reduce the SHGC to 0.24, limiting unwanted heat gain during the summer.

Heat-Absorbing Surface and Thermal Mass: High-mass floors provide the heat-absorbing surface and thermal mass for both levels of this two-story house. The house rests on a dark-pigmented concrete slab, insulated from the ground on all sides by 4 inches of extruded polystyrene foam. On a sunny winter day, the 5-inch-thick slab, which serves as the finished floor for the main level, absorbs solar energy, warming slowly due to its thermal mass and releasing heat gradually during the night. The exposed area of the slab is greater than 500 square feet, more than enough to absorb and store heat from the 75 square feet of window glazing, preventing overheating even on the sunniest days.

The upstairs floor consists of locally milled ¾-inch-thick solid oak boards made from trees harvested from the building site. The owners chose to remove these trees to increase solar access. To improve the heat-storing capacity of the wood floors, the owners packed 1½ inches of clean sand beneath the boards. An exposed floor area of over 400 square feet adequately stores solar energy from the 63 square feet of window area.

A living roof composed of native drought-tolerant plants replaces conventional roofing above the main floor and minimizes reflection of summer light through the upstairs windows, limiting unwanted heat gain. Transpiration and evaporation from the living roof provide cool, fresh air for open windows. In the winter, the living roof holds snow that reflects the low rays of the sun through the windows and increases solar gain when it is most needed.

Heat Distribution: An open floor plan on the main floor allows heat to pass freely from room to room, and the lightly colored walls and ceiling reflect heat and light throughout the building. Similarly, the upstairs walls are lightly colored, and the vaulted ceilings and open spaces allow easy distribution of heat. Energy efficient ceiling fans offer an option for additional cooling and heat distribution when necessary.

Control Measures: Roof eaves extending 2½ feet beyond the walls on both the upper and the lower levels act as controls. Deciduous trees on the east, west, and north sides of the building further reduce summer heat gain.

Performance: From April through October, the home remains comfortable without supplemental heating or cooling, relying exclusively on passive solar design. During the winter months, when cold temperatures increase heating requirements and prolonged cloudiness reduces passive solar gain, a high-efficiency woodstove adds supplemental heat. Depending on the cloudiness and severity of the winter, the stove burns 1.5–2 cords of wood per year. Propane provides fuel for cooking, consuming 40–50 gallons per year, and electricity operates a refrigerator, freezer, and other household appliances, amounting to 2000 kWh per year.

CASE STUDY 3.2

High Mass House

Location: Tucson, Arizona (32.1° N latitude)

Year built: 1998

Size: 2240 ft²

Climate: Tucson experiences mild winters; hot, dry summers; and a moderate climate during spring and fall. Tucson is surrounded by desert, receives only 10–12 inches of rain per year, and has nearly 200 cloud-free days annually. In this climate, summer cooling is the dominant concern. Nevertheless, buildings must also have adequate insulation and thermal mass to moderate daily temperatures swings and keep living spaces comfortable all year.

System Design Preference: All three passive solar systems—direct gain, thermal storage walls, and thermosiphon systems—could work well in this climate. However, the size and shape of the building site prevented the house from being optimally oriented, prompting the designer to select active systems to collect and distribute heat.

Solar Collector and Heat Distribution: Three flat-plate solar hot water collectors mounted on

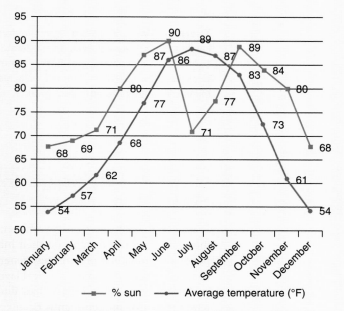

The average monthly temperature and percentage of sunshine for Tucson, Arizona, indicate that some form of cooling is required most of the year. The cloudiest month in Tucson provides more sunshine on average than the clearest month in northern Wisconsin. © 2016 Cengage Learning®.

Small windows and ample thermal mass minimize summer overheating in this home, which is accessible to persons with disabilities. Courtesy of Scott Grinnell.

the roof provide heat for the house when necessary. Water flows through the solar collectors, warms to temperatures of 160°F or more, and passes through a heat exchanger in a large insulated tank. Electric pumps circulate hot water through pipes in the floor, which radiate heat throughout the house.

This house also employs two innovative low-energy cooling systems. The first system uses rooftop radiators made of corrugated steel, which cool quickly to the night sky. High-efficiency pumps circulate fluid from the radiators to the interior masonry walls, extracting heat from the walls and cooling the house by as much as 7°F–8°F during the night. The system shuts down during the day or when nighttime cloud cover prevents adequate cooling. The second system uses a variable-speed blower to draw cool morning air into the house. When needed, a small pump saturates a screen near the blower and moistens the air. Evaporation further cools the air and can lower the temperature within the house by as much as 25°F. The system uses 40–70 gallons of water per day, drawn from 4000 gallons of collected rainwater.

The few windows in the house are small and do not serve as significant solar apertures. All windows are coated to minimize heat gain, achieving an SHGC of 0.24 and a U-factor of 0.29.

Thermal Mass: Insulated concrete forms comprise the external walls, providing both insulation and thermal mass to moderate daily temperature swings and keep living spaces comfortable all year. Plaster further contributes to the thermal mass of the walls. Tumbled brick used for floors and internal walls also adds substantial thermal mass to the structure, minimizing interior temperature swings.

Internal Spaces in a Passive Solar Home

Passive solar homes are energy-efficient, highly insulated structures designed to make internal spaces comfortable for their inhabitants. While the specific layout of rooms depends on individual preferences, general design principles help direct the best use of spaces. Figure 3.19 depicts a typical layout for direct gain passive solar homes.

For direct gain homes, spaces illuminated by sunlight are bright and warm and tend to be those predominantly occupied by people during the day. These include living rooms, dining rooms, and some craft or hobby rooms and should be placed on the south side of the house. Bedrooms, which are not heavily used during the day and typically do not need heat until evening, can be placed away from direct gain locations or on the east, west, or north sides of the house. Whether bedrooms receive morning or evening light

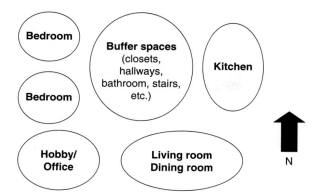

FIGURE 3.19 A direct gain passive solar home typically arranges spaces occupied during the day on the south side, while those that produce their own heat or are occupied at night belong on the north side.
© 2016 Cengage Learning®.

depends on occupant preference. Bathrooms, closets, storage rooms, and stairways can fill in the remaining spaces in the center of the house. Spaces that generate their own heat, such as kitchens and utility rooms, may also be placed on the north side.

Chapter Summary

Elements of passive solar design date back thousands of years. Only relatively recently, with the availability of inexpensive fossil fuels, have builders ignored the value of solar design. Yet many passive solar features cost no more to build than conventional construction. With no additional operating costs, passive solar design provides free energy for the life of the building, reduces adverse impacts on the environment, provides greater economic security, and creates more comfortable living spaces. Passive solar design is an extremely valuable component of green building.

Review Questions

1. Limited windows on the north, east, and west sides of a solar home can result in dim internal conditions that encourage occupants to use artificial lighting, eroding some of the energy savings inherent to passive solar design. Without increasing the number or size of windows, what other approaches can make internal spaces brighter?

2. An architect has designed a glass-covered pond for the rooftop of an office building in southern New Mexico. In the winter, the pond fluid (a mixture of water and antifreeze) absorbs sunlight during the day, and pumps circulate the warm fluid throughout the building during the night. Automatic temperature-controlled motors draw insulation panels over the pond at sunset, preventing heat loss to the night sky. In the summer, the insulation panels cover the pond during the day and are withdrawn only after sunset to allow the fluid to cool to the night sky. Pumps circulate the cool fluid throughout the building during the day. Is this a passive solar system? Assess its effectiveness. Is it a good idea?

3. A well-insulated wall separates a solarium from an attached house. A designer is considering two different methods to regulate heat flow into the house. The first method will install thermostat-controlled vents between the solarium and the house, automatically opening and closing the vents, depending on the temperatures. The second method will rely on several large, manually operated windows to allow solarium heat to enter the building. Which would you recommend, and why?

4. The second story of the northern Wisconsin home in Case Study 3.1 uses ¾-inch-thick solid wood flooring over a 1½-inch-thick sand bed in a direct gain passive solar design. Identify the absorber, thermal mass, and predominant heat distribution method for this system. Discuss the effectiveness of this system. How do you think it compares in performance to the solid concrete slab of the first floor?

5. Windows that serve as solar collectors in direct gain passive solar homes should be as transparent as possible to maximize solar heat gain. However, most triple-pane windows with low-E coatings offer an SHGC of only 0.57 or less (by comparison, a single pane of clear glass has an SHGC of 0.78). Multiple panes of glass (which result in greater reflection) and low-E coatings (which block infrared wavelengths) both act to limit the transmittance. For a home that requires maximum solar heat gain, what arrangement can be made without excessively compromising insulation?

6. An architect designs a theatre to harness solar energy without the glare of a direct gain system, as shown in Figure 3.20. The insulated south-facing wall includes vents at the bottom and top. Outside the wall, a black metal absorber plate heats quickly in the sunlight behind a large double-pane window. Temperature-controlled fans draw the warmed air into the theatre, which is replaced by cool air through the bottom vent. At night, the vents are closed to prevent heat loss. Compare this system to a Trombe wall. How is it similar? How is it different? How would you classify this type of system?

7. Most entry doors allow more air infiltration than high-quality awning or casement windows. What design options can help mitigate this problem?

8. A builder in Miami, Florida, is interested in passive solar design. Miami only rarely experiences cold weather. The average temperatures for January and July are 68°F and 84°F, respectively, with high humidity year round. Conventional buildings rely on air conditioning and dehumidifiers to improve building comfort. What lower-energy solutions would you recommend? What should be the primary focus of the solar design?

9. In 2003, the Hartley Nature Center in Duluth, Minnesota, installed a black metal solar collector on the south wall of its visitor center (Figure 3.21). On a sunny day, fresh air flows

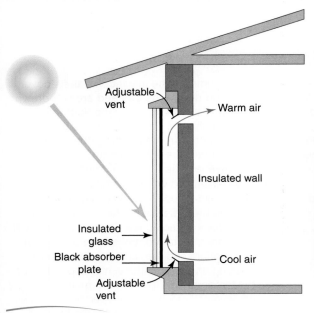

FIGURE 3.20 Sketch of wall. © 2016 Cengage Learning®.

FIGURE 3.21 Solar Wall. Courtesy of Scott Grinnell.

through tiny holes in the metal baffle, heats rapidly in the sunlight-warmed collector, and enters the building through ducts at the top of the collector. A variable-speed fan distributes heat throughout the building. Temperature

sensors operate dampers to prevent the system from operating when too cold or too warm. How would you classify this type of system? Is it passive or active? How is this system different from the one in Question 6?

Practice Problems

1. An artist living in Maine wishes to use passive solar design to heat a small studio that is separate from the main house. The artist wishes the studio to be illuminated only by indirect northern light to avoid glare. Design the structure with appropriate passive solar features. Consider the orientation of the studio, construction materials, and type of system.

2. A Trombe wall, if correctly vented, can double as a solar chimney for summer cooling. Sketch a building design that implements such a feature.

3. Consider a one-story, two-bedroom, one-bath home with a simple rectangular footprint measuring 30 feet × 40 feet. Sketch the general locations of the following rooms: bedrooms, kitchen, dining room, living room, bathroom, laundry room, pantry, and three closets. Indicate the orientation of the building, the type of solar system selected, and the climate for which the building is designed.

4. In regions that experience extended cloudiness, most buildings designed to harness solar energy still require auxiliary heat. The Isabella Experiment Station in northern Minnesota is a 2100-square-foot house that can be heated entirely by solar energy. In addition to a direct gain passive solar design, the superinsulated house uses solar hot water collectors to absorb and store enough solar energy during the summer to heat the house all year. Beneath the house is a basement filled with 210 cubic yards of sand and taconite pellets insulated by 16 inches of expanded polystyrene rigid foam. During the summer, pumps circulate hot water through

pipes embedded in the sand and taconite, and the stored energy can be subsequently withdrawn at any time. Assume the summer sun can warm the entire volume of sand and taconite to a temperature of 160°F and that the soil temperature around the basement is 40°F. The total area of the basement walls and floor is 1800 square feet.
a. At what rate is heat conducting into the ground through the basement walls and floor?
b. If the sand and taconite pellets can store a total of 12 million Btu of thermal energy, what percentage of the thermal energy would be lost each month if left unused? Assume the temperatures remain constant.

5. On a cold, sunny winter day, a particular passive solar home absorbs 200,000 Btu of solar energy. During the following 24-hour period, the same home loses 260,000 Btu of heat by conduction, convection, and radiation. The windows account for 92,000 Btu of that loss. The owners discover that by installing window quilts they can prevent a loss of 30,000 Btu on winter nights.
a. What percentage of savings does this represent for the owners' conventional heating system demands?
b. If these conditions remain constant and the heating bills are $200/month, how much will they save each month?
c. If a set of window quilts cost $2200, how many winter months will it take to break even?

6. A builder of a 2000-square-foot passive solar home in Fairbanks, Alaska, wishes to install sufficient south-facing windows to completely

heat the home in April, when the average temperature is 29°F and the home requires 210,000 Btu of heat.

a. What total surface area of window glass is required to meet the heating needs if, on average, 3000 Btu of solar energy enter the home for every square foot of window glass?

b. In January, when the average temperature is −12°F, the house requires 420,000 Btu of heat per day. What fraction of the total heating requirements will the windows meet if the January sun offers only 450 Btu/ft^2?

c. Give at least two reasons why designing for the average April day is an unreasonable approach. What would be better?

Alternative Construction

Courtesy of Kelly Hart.

Introduction

Mud and other natural materials have shaped human shelters for more than 10,000 years. Across vastly different climates, people have created structures using little more than the soil beneath their feet and the plants that grew on it. Builders shaped majestic mosques from mud bricks, assembled plateau-top cities of adobe (Figure 3.12), erected sections of the Great Wall of China with rammed earth, constructed cottages made of cob (Figure 4.1), and raised school buildings and churches from straw (Figure 4.2), all with practical techniques that have proven durable. For regions that lack millable timber, alternative forms of construction offer simplicity, economy, and self-reliance.

FIGURE 4.1 Built in 1865, this cob cottage once housed a family of early pioneers in the temperate climate of Marlborough, New Zealand. Created from local materials, the original cottage included 16-inch-thick cob walls, rammed-earth floors topped with a thin layer of concrete, and timber cut from dead and weathered pine. The building served as a schoolhouse between 1906 and 1909 and is now operated as a museum by the Marlborough Historical Society. Courtesy of Francis Vallance.

FIGURE 4.2 The scarcity of timber and the presence of poor-quality sod in the Sandhills region of Nebraska led to the use of straw bales as a construction material not only for homes and school buildings but also for this church built in the town of Arthur in 1928. Courtesy of the Nebraska State Historical Society.

The use of earthen and other natural materials flourished for centuries. The particular techniques and designs varied from place to place, depending on climate and available resources. This changed dramatically after the Industrial Revolution, particularly for Europe and America, when standardized manufactured products offered reductions in on-site labor. Long-distance transportation resulted in a greater choice of materials, and because of its simplicity and efficiency, building with standardized dimensional timber gradually replaced traditional techniques. Logging became a major industry, and vast areas of wild forests fell to clear-cutting machines. Over a relatively brief period, loggers claimed most of the easily accessible old-growth timber.

Awareness of the ever-increasing stress on the world's remaining forests prompts renewed interest in these traditional and economical building techniques, known today as alternative construction. Using natural or recycled materials, alternative construction allows builders to create comfortable spaces with greatly reduced impact on the environment. By making use of local materials available at no or very little cost, alternative construction can be less expensive than conventional timber construction. What is not free is the increased labor, which can be substantial. Even when supplied by the owner, the cost of labor cannot be ignored. On the other hand, this labor can be beneficial by uniting groups in a common cause and prompting social interaction and community involvement. However, these social benefits, as well as any financial savings, disappear if the labor is hired out to professional construction companies.

Most alternative construction techniques bear evidence of the human touch in the finished product (Figure 4.3). In conventional construction, machines mill and manufacture the building materials—dimensional studs, plywood, drywall, and so on—producing a finished structure that can be uniform and sterile. The older techniques were originally chosen out of necessity, utilizing resources readily available. Now, many turn to alternative construction as a means to surround themselves with spaces that feel handmade and natural. Creating a comfortable ambience—however individuals choose to define this—is an important aspect of green building.

FIGURE 4.3 The ability to customize homes and create spaces that embody the human touch inspires many to turn to alternative building styles, such as this cob structure built in Vancouver, British Columbia. Courtesy of David Sheen.

This chapter briefly summarizes some of the most common forms of alternative construction, examining the benefits and limitations with regard to the principles of green building. All of the techniques can utilize passive solar design to reduce energy consumption and improve occupant comfort.

Common Features

All forms of alternative construction make use of local, readily obtainable materials. Many of these materials consist of waste products from other processes, such as straw from grain production, scrap wood from logging, and used automobile tires from landfills. Others utilize natural materials obtained from the building site, such as clay and sand. Consequently, a major benefit of alternative construction is that it achieves very low embodied energy through minimal material processing and reduced transportation.

Another common characteristic of alternative construction is that it can generally be done with relatively unskilled labor and with little use of machinery. This presents greater opportunities for owner involvement than conventional construction but also requires a larger investment of labor. Nonetheless, for many, the ability to customize spaces—to build a structure rather than have one built—justifies the extra labor and makes alternative construction appealing.

Earthen materials like adobe, cob, and rammed earth tend to have poor insulation value. To create comfortable spaces with these materials, structures often employ very thick walls. While the extra thickness does increase the insulation value somewhat, it is often not enough for cold climate construction. However, the extra thickness substantially adds to the structure's thermal mass, making these materials well suited to relatively mild climates, where moderating temperatures with high thermal mass is more practical than minimizing heat flow with high levels of insulation.

The thick walls of these structures take away from the available living space, however, leaving less of the building's footprint usable for occupation. This requires larger buildings (and hence more disturbance to the environment) for the same internal space offered by conventionally built structures.

Adobe

Adobe is one of the oldest and most widely used building materials. Consisting of sand, clay, and a fibrous material like straw mixed together with water, adobe is formed into blocks and dried in the sunlight (Figure 4.4). The dried blocks provide a fireproof, nontoxic building material that can be mortared together with a layer of mud applied between the blocks. While adobe structures can be built in many regions, the technique is most widely used in arid climates, where blocks dry quickly.

Adobe is natural, readily available, and inexpensive and requires no special equipment to use. When the blocks are made by hand, as is traditional, adobe has very low embodied energy. Machines are now available that mix adobe and form blocks, increasing the embodied energy considerably. Adobe offers excellent thermal mass but provides very little insulation (R-value = 0.2/inch – 0.5/inch), making it

FIGURE 4.4 Sun-dried adobe blocks are an ancient building material that continues to be used around the world. When mixed and formed by hand, locally sourced adobe creates a durable building material with very little embodied energy. The blocks shown provide structure for Hummingbird Studios in southwestern Oregon. Courtesy of Leslie Lee.

best suited for mild climates, particularly those with large daily temperature swings, where its thermal mass can delay solar heating much like a Trombe wall. Historically, adobe walls varied in thickness from 10 to 30 inches, and the thicker walls tended to moderate temperatures over periods of weeks or months, creating an internal temperature similar to the average monthly temperature (Figure 4.5).

Adobe structures can be extremely durable and account for some of the oldest buildings in the world (Figure 4.6), some dating back more than a thousand years. However, the high mass and brittle behavior of adobe-built structures cause them to perform poorly during earthquakes, making them unsafe for seismically active areas. Reinforcement of adobe structures with metal or wood can greatly improve their performance, but most traditional adobe structures lack reinforcement.

Adobe-built structures require a solid, well-drained foundation to prevent the massive walls from settling and cracking. The foundation must also protect the adobe from groundwater and precipitation.

FIGURE 4.5 Traditional adobe construction, shown here in the remnants of a wall at Mission Santa Clara in California, offers enormous thermal mass but relatively poor insulation. Courtesy of Charles Berry, Santa Clara University.

FIGURE 4.6 This traditional adobe house, originally built in the 1850s in Las Vegas, New Mexico, includes a north-opening courtyard that shades occupants throughout the day. Courtesy of Scott Grinnell.

Cob

Cob is a building material as ancient as adobe, also consisting of sand, clay, and straw, though in slightly different proportions. Unlike adobe structures, which consist of stacked bricks, cob construction creates a single, seamless entity by working layers together (Figure 4.7). Cob also differs from adobe by having a larger proportion of straw, which improves its insulation value somewhat.

FIGURE 4.7 Supported by a stone foundation, 18-inch-thick cob walls dry in the Iowa sun and frame the beginning of a 300-square-foot home. Built layer by layer from local materials, the cob walls allow for hand-sculpted spaces and custom designs. Courtesy of Hap and Lin Mullenneaux.

FIGURE 4.8 Constructed in 2008 for roughly $7000 and with 1500 hours of labor, this cob home is nearly finished and ready for the final coats of plaster. Courtesy of Hap and Lin Mullenneaux.

FIGURE 4.9 Lime plaster gives this Iowa home a white appearance and allows the cob walls to breathe. The simple roof design and large eaves direct rainwater away from the walls and foundation. Gutters collect some of the water for on-site usage. Courtesy of Hap and Lin Mullenneaux.

Cob construction mixes sand, clay, and straw with water—traditionally by foot or with the assistance of horse or ox—to create a sticky, malleable material that can be sculpted by hand to create thick, rounded walls and smoothly curving passageways. Cob is fireproof and resistant to cracking, making it useful in many different regions around the world. When obtained locally and mixed by hand, cob is inexpensive and has very low embodied energy. It is easy to use, requires no special training or machinery, and can take on a variety of textures and forms (Figure 4.8).

Cob has poor insulating value (R-value = 0.3/inch − 1.0/inch), so it is best suited for use in relatively mild climates. In the United Kingdom, where cob construction has a long history, old buildings have walls that are 30 or more inches thick. Such a wall could provide sufficient insulation (up to R-30) to keep the inside warm during the winter. More importantly, its enormous thermal mass would maintain an extraordinarily stable interior temperature, changing only slightly over the span of months.

Although the building materials are natural and inexpensive, cob construction is labor intensive, and cob-built structures benefit from the assistance of many workers. Cob dries to be nearly as hard as concrete and can be covered with plaster or stucco to reduce damage from water. Like adobe, cob construction requires a solid foundation to prevent settling, and must be kept dry and protected from rain and groundwater (Figure 4.9).

Cordwood

Cordwood construction, another old building technique, makes use of short logs, usually 12 to 24 inches in length. Stacked like a row of firewood, each log has one end exposed on the interior surface and the other protruding from the exterior (Figure 4.10). Each course of logs is pressed onto a bed of mortar, which can be cob, wet lime, or various

FIGURE 4.10 Two 3-inch strips of cement mortar hold 12-inch sections of locally cut cedar and pine in this cordwood wall. Dry cedar sawdust fills the 4-inch space between the mortar strips and provides insulation. Courtesy of John Olson.

FIGURE 4.11 Cordwood construction adds natural beauty to this home in northern Wisconsin. Using local and recycled building materials, homes like this cost only a fraction of conventional construction. Courtesy of Scott Grinnell.

mixtures of concrete. With the log ends left exposed, variations in the diameter, shape, and growth rings contribute to the aesthetics of this building technique.

Historically, the logs used in cordwood construction were leftovers from the logging industry: those unsuited for sawmills or too crooked for log cabins. Due to its low cost, cordwood construction became popular during the Great Depression, particularly in Wisconsin's logged-over areas where plentiful scrap wood provided ample raw materials (Figure 4.11). The use of scrap wood gives cordwood construction very low embodied energy, especially when securing the logs with natural earthen mortar. While labor intensive, cordwood construction has the advantage of being manageable by a single person working alone: much like stacking firewood, a single person can easily heft and position each piece.

Cordwood construction generally offers better insulation than pure cob. Two strips of mortar, on the inside and outside edges of the wall, are generally adequate to anchor each row of logs, allowing space for insulation around the logs near the center of the wall. Furthermore, mortar can have additives, such as paper, straw, wood chips, and pumice, depending on what is available, which further improve the insulation value.

A factor limiting the insulation of cordwood construction is the transverse arrangement of logs, where one end of the log is inside the structure and the other end is outside. This orientation allows greater heat conduction through the wood than traditional log construction because heat flows more easily with the grain than across it. For dry cedar, the R-value is roughly 1.5/inch across the grain but as low as 0.5/inch with the grain. Cordwood construction yields an R-value of roughly 1.0/inch to 1.5/inch, depending on the species of wood, cavity insulation, and proficiency of the builder. A 24-inch-thick cordwood wall can yield an insulation value as high as R-36, which is adequate even for cold climates.

Thick walls of wood and mortar provide substantial thermal mass that moderates temperature swings and adds to the comfort of the structure. However, compared to abode and cob construction, cordwood offers less thermal mass for the same wall thickness. Additionally, the thermal mass is less continuous because of the insulation between the mortar strips, so walls made of cordwood do not offer the same Trombe wall effect as adobe and cob.

Adequate drying, debarking, preparation, and selection of logs are critical in cordwood construction because they can reduce cracking and gapping around the logs. Wood naturally expands and contracts with changes in humidity and temperature. Humid summer days cause wood to expand, which can lead to cracks in the mortar and structural damage to the building. Dry winter days cause wood to shrink, which can produce gaps between the logs and the mortar, permitting unwanted infiltration of air, insects, and vermin, compromising comfort and the insulation value. Gaps can be remedied in part by caulking around the logs once they have dried, but this further adds to the labor of the process. To minimize these problems, cordwood construction benefits from meticulous attention to detail, particularly in extreme climates.

Earthbag

Sacks filled with sand have a long history of use: to control flooding, to minimize erosion, to serve as military bunkers, to stop bullets in shooting ranges, and so on. Versatile and easy to use, earth-filled bags can also form walls and domed roofs to become buildings of free-form architecture. Earthbag construction became popular in the United States in the 1970s as a means to create inexpensive shelters capable of withstanding severe weather and earthquakes.

Earthbag construction requires sturdy sacks, such as polypropylene feed sacks, and a source of inorganic earth, which includes most combinations of clay, silt, sand, and gravel. The range of suitable material for earthbag construction offers a significant advantage over adobe and cob, both of which require specific mixtures of sand, clay and straw.

Earth-filled bags placed on top of each other, row by row, create thick, high-mass walls. Modern earthbag builders insert strands of barbed wire between each row to prevent the bags from slipping or shifting. The barbed wire grips the polypropylene sacks so effectively that earthbag structures are often able to withstand hurricanes and earthquakes. Earthbag construction is also fireproof, resistant to flood damage, and unappealing to termites.

Earthbag construction offers the added benefit of being able to serve as its own foundation, provided the bags in contact with the earth are filled with gravel or other coarse material that prevents moisture from wicking upward. In climates that experience freezing conditions, the foundation must be well drained or insulated (or both) to prevent structural damage from frost heave.

Like other forms of earthen construction, most earthbag structures have enormous thermal mass but poor insulation value, creating a Trombe wall effect that is best suited to mild climates with large daily temperature variations. However, earthbag construction can also be adapted to cold climates by filling the bags with insulating materials, such as vermiculite, rice hulls, or volcanic scoria. The material must be nearly incompressible to prevent settling and shifting (Figure 4.12 and Figure 4.13).

The components for earthbag construction—feed sacks, barbed wire, and soil—are inexpensive and broadly available, making this building technique extremely versatile.

FIGURE 4.12 Polypropylene rice sacks filled with volcanic scoria form the walls of this passive solar home in Colorado. Thick plaster made from shredded paper and cement smooths gaps in the bags and contributes to the overall insulation, which is sufficient to keep this house warm in the winter even at its 8000-foot elevation. Courtesy of Kelly Hart.

FIGURE 4.13 The exterior buttresses supply lateral support to the walls. Although earthbag construction can be applied to nearly any design, it is most stable when employing curved architecture. Courtesy of Kelly Hart.

When soil is obtained from the building site, earthbag construction offers low embodied energy. It is labor intensive but requires no special machinery or training. However, polypropylene sacks must be protected from sunlight to prevent degradation and are typically coated with plaster or stucco as soon as feasible.

Rammed-Earth Tire

Rammed-earth tire structures utilize discarded automobile tires packed tightly with soil and stacked row upon row around a partially subterranean passive solar home. First popularized in New Mexico as **Earthships** by Michael Reynolds in the 1970s, these structures grew out of a philosophy of living harmoniously with nature and making use of locally available by-products of society (see *Further Learning: Earthship Philosophy*). The thousands of waste tires crowding landfills across the country became a valuable building resource (Figure 4.14). Removing these tires from landfills simultaneously mitigated environmental and human health concerns associated with their role as mosquito-breeding grounds and as sources of toxic runoff and air pollution caused from burning. Earthship construction was seen as a means of reducing some of society's enormous disposal problems.

Further Learning EARTHSHIP PHILOSOPHY

The **Earthship philosophy** promotes the use of locally available materials that require little or no additional energy input, are durable, and can be assembled with minimal training. These materials may be either natural or recycled from waste generated by society. The philosophy promotes simplicity, self-reliance, and the use of natural energy sources, such as solar and wind. Earthships typically collect rainwater for domestic use, incorporate greenhouses for food production, treat sewage on site, utilize natural heating and cooling methods, and remain disconnected from the utility grid.

FIGURE 4.14 Discarded tires comprise the basic structure of Earthships. Workers laboriously pack each tire with earth using sledgehammers. Courtesy of Monica Holy.

Rammed-earth tire structures are massive, as each tire packed full of soil weighs over 300 pounds. The total mass of thousands of packed tires is amplified by being partially buried, usually in a south-facing hillside. Authentic Earthship construction leaves the walls and floors uninsulated with respect to the ground, creating an interior temperature strongly influenced by that of the ground itself. At sufficient depth, the ground temperature remains nearly constant at the annual average surface temperature. For climates with abundant winter sunshine, large south-facing windows capture sufficient sunlight to maintain a comfortable interior temperature (Figure 4.15). The higher summer sunshine enters less of the building, allowing the influence of the ground to keep the structure cool. For climates that commonly experience extended winter cloudiness, however, rammed-earth tire structures require insulation around the perimeter walls to avoid excessive heat loss. Window quilts or similar insulating measures also improve comfort.

Rammed-earth tire construction is extremely labor intensive, requiring excavation of the site, sledgehammers to pack soil, and the transportation and handling of thousands of tires. In addition, these structures are limited in design due to the large size and unvarying shape of the tires. Modern variations on the original Earthship design now incorporate a variety of techniques, including cob and straw-bale construction, that increase design flexibility considerably. Furthermore, structures designed for climates with frequent winter cloud cover use various strategies for heat storage and insulation, as discussed in *Further Learning: Passive Annual Heat Storage.*

Although automobile tires remain plentiful in landfills, more than 70% are now recycled into new products. These include rubber-modified asphalt, athletic and playground surfaces, insulation, roofing materials, conveyor belts, and hazardous waste containers. Furthermore, tires are a valuable fuel source, as they are made of energy-rich petroleum. Consequently, the benefits of salvaging old tires for constructing homes may no longer be as compelling as when Earthship construction began in the 1970s, particularly with prospects of improved recycling. Nevertheless, the Earthship philosophy remains a powerful attraction to many who seek a simple, independent, and low-impact lifestyle.

FIGURE 4.15 Largely buried, Earthships moderate daily and seasonal temperature variations through enormous thermal mass and large south-facing windows. This Earthship, constructed under the guidance of Michael Reynolds in 2011, is located on Gulf Islands, British Columbia. Courtesy of Monica Holy.

Further Learning PASSIVE ANNUAL HEAT STORAGE

In some cold climate locations, persistent cloud cover during winter months renders conventional passive solar design ineffective. The control measures appropriate to conventional passive solar design—maximizing solar heat gain in the winter, when needed, but minimizing it in the summer, when not needed—fail to deliver the much-needed winter warmth. Without sunshine, large south-facing windows simply serve as sources of net heat loss for the building.

An alternative strategy, called passive annual heat storage, uses earth-sheltered structures with unshaded south-facing windows that absorb solar energy at all times of the year, particularly during the summer. While this would significantly overheat ordinary buildings, specially designed earth-sheltered structures apply the enormous thermal mass of the ground beneath and surrounding the structures to maintain a relatively stable interior temperature year-round.

Passive annual heat storage relies on the relatively poor insulating value of soil to transfer heat from the building into the soil during periods of excess solar heating. A large area of buried insulation—an umbrella above the structure and extending outward from the periphery walls 20 feet or more—reduces heat lost to the surface environment. Thermal energy trapped beneath the umbrella remains in the soil, stored as if in an enormous battery.

These earth-sheltered structures are typically made of concrete and completely buried except for south-facing windows. Several feet of soil cover the structure's uninsulated roof, above which the umbrella is capped with several feet of additional soil. The umbrella, which might be composed of 4 inches of extruded polystyrene covered by polyethylene sheathing, not only provides insulation but also shields the structure from moisture. Appropriate slopes and drains allow for storm water runoff. As long as the soil beneath the umbrella remains dry, its enormous volume retains much of the thermal energy absorbed during the summer and makes this energy available during the winter as the building cools relative to the surrounding soil.

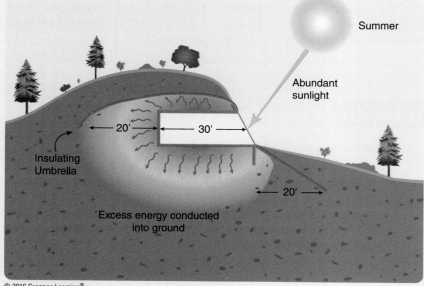

Summer

Abundant
sunlight

← 20' → ← 30' →

Insulating
Umbrella

← 20' →

Excess energy conducted
into ground

© 2016 Cengage Learning®.

(continues)

(continued)

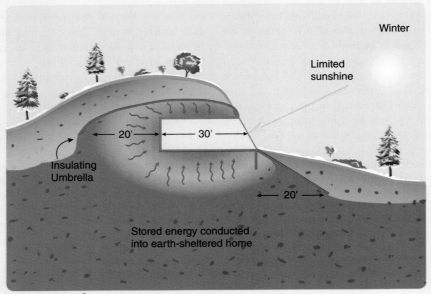

Winter

Limited
sunshine

20'

30'

Insulating
Umbrella

20'

Stored energy conducted
into earth-sheltered home

© 2016 Cengage Learning®.

Although the R-value for dry soil varies with composition, a rough value of R-2 to R-4 per foot provides an overall resistance to conductive heat flow of R-40 to R-80 over the 20-foot protected area on each side of the building. This R-value limits the amount of heat conducted beyond the 20-foot protected area and makes a significant portion still available to the earth-sheltered structure through the winter. These structures might, therefore, remain comfortable year-round without *any* supplemental heat, even in locations lacking winter sunshine.

CASE STUDY 4.1

Rammed-Earth Tire House

Location: Colorado (39° N latitude)

Year built: 2001

Size: 1950 ft²

Cost: $156,000 ($80/ft²)

Description: Located in the Rocky Mountains at an elevation of 8400 feet, this partially submerged passive solar home relies on earth-filled automobile tires for much of its structure.

Benefitting from more than 250 sunny days per year, this cold climate home approaches net-zero energy consumption by generating its own electricity with a solar-electric system (also described in Case Study 7.4 of Chapter 7).

A total of 950 used automobile tires came from a tire company located 30 miles from the construction site. It took the owners 10 weeks and about 1000 labor-hours to pack the tires with 500,000 pounds of decomposed granite excavated from the site and to arrange the tires into four bays stacked 10 courses high. Cob fills the voids between tires and finishes the interior walls. The combination provides enormous interior thermal mass that greatly moderates temperature swings. The tire walls rest directly on the ground, which is kept dry by surrounding drain tiles. The curved shape of the four bays and the interlocking placement of the tires give the structure great stability. The tires, therefore, serve as their own foundation.

Four U-shaped bays of tires filled with decomposed granite comprise the basic shape and structure of this home. The top course, not yet installed, will include bolts anchored in concrete to secure roof framing. Courtesy of Jerry and Diana Unruh.

The buried portion of the walls, as well as the 5-inch-thick concrete slab poured around the walls, is not insulated from the ground. This lack of insulation enables the ground—which remains year-round at roughly 55°F beneath the floor—to strongly influence the interior temperature of the home. However, passive solar design and abundant sunshine warm the interior to comfortable levels during the winter, while permitting the ground to cool the home during the summer. The combination of passive solar design and thermal contact with the ground represents a year-round compromise that works well with the climate.

Standard dimensional lumber frames the roof and remaining wall sections, and fiberglass batts provide an insulation value of R-38 in the ceiling,

The completed home enjoys abundant sunshine and a view of mountain peaks. Courtesy of Michael Shealy.

R-42 in the aboveground side walls, and R-26 in the aboveground rear wall. A standing-seam metal roof includes five operable skylights that bring light and fresh air into the home. A total of 410 square feet of south-facing double-paned windows, including six operable windows at the base of the fixed windows, gathers sunlight and allows for ventilation when necessary. Window shades offer an added insulation value of R-2 at night.

The large south-facing windows promote generous daytime heating and can raise the interior temperature above 80°F on a sunny winter day. The building's thermal mass retains this warmth throughout the night, and even after three consecutive days of cold, overcast weather—which at this site is a rarity—it prevents the house from dropping below 55°F without supplemental heat.

Tile flooring absorbs solar energy and gradually releases heat throughout the night. Cob covers the tire wall in the foreground and adds interior thermal mass, contributing to a uniform and comfortable temperature. Courtesy of Michael Shealy.

Although the sun provides an estimated 90% of the building's heating requirements, a backup propane log stove adds supplemental heat when necessary. Propane also provides fuel for cooking, heating water, and drying clothes, using a combined total of roughly 300 gallons per year.

The owners contracted out the electrical, plumbing, and framing work but otherwise did most of the labor themselves. By investing more than 3100 hours into the construction of their home, they saved an estimated $80,000 in labor costs. Perhaps more importantly, their labor allowed them to add personal touches, create custom spaces, and design a long-lived, climate-appropriate home that suits their needs.

Straw Bale

Using bales of straw to construct buildings began in the late 1800s on the prairies of Nebraska as an alternative to sod houses. In areas that lacked plentiful timber, bales of straw provided a means to use local resources as a building material and create structures quickly and inexpensively (see Figure 4.2).

Straw is an agricultural waste product left over from cultivating grains, such as wheat, oats, and rye. After removing the grain, farmers traditionally gathered the straw for use as animal bedding, as garden mulch, or for erosion control, or they simply left it on the fields and burned it after harvest.

FIGURE 4.16 Straw bales fill around the timber frame structure of this two-story passive solar home in northern Wisconsin. Conventional stud framing supports windows and doors. Courtesy of Greg Weiss.

Modern straw-bale construction creates walls with tightly bound bales of dry straw, stacked in rows and tied together, as shown in Figure 4.16. A conventional roof either is supported by timbers that bear the weight or rests directly on the bales. The use of supporting timbers is generally favored in all but the driest climates because it allows the installation of the roof prior to working with straw. The protection offered by the roof reduces the risk of the bales getting wet during construction. Wet straw offers a welcome environment for mold, fungi, and insects and leads to degradation that severely compromises the structure.

Straw can be sustainably grown in a short period of time. Locally harvested straw represents a renewable, nontoxic, and biodegradable resource with very low embodied energy. Straw also has fairly good insulating properties, yielding an R-value of 2.4/inch or more. However, in straw-bale construction, this value is typically closer to 2.0/inch due to air movement between bales. Nevertheless, with straw bales varying in thickness from 18 to 24 inches, the insulation value for a straw bale wall becomes R-36 to R-48, suitable even for very cold climates.

Although straw is much less dense than earthen walls, the thickness of the bales creates enough thermal mass to moderate interior temperatures, adding to the comfort of straw-bale buildings. Like structures using other thick-walled building techniques, however, straw-bale structures require larger footprints for the same size interior space compared to conventional construction.

Straw-bale construction has proven itself durable—with early structures dating back more than a hundred years—but this is true only if the straw is kept dry. To accomplish this, the straw bales must be separated from the foundation by a moisture barrier, sealed behind stucco or plaster, and protected from weather by large roof eaves, proper drainage around the building, and generous external windowsills that direct water away from the walls (Figure 4.17 and Figure 4.18).

Straw-bale construction is considerably less labor intensive than earthen techniques, particularly rammed-earth tire construction, since the bales are much less dense and more easily managed. Straw-bale construction lends itself to great flexibility of design, requires relatively little training, and utilizes no special machinery.

FIGURE 4.17 Concrete stucco applied over 1-inch hex wire netting seals and protects the straw from weather. Courtesy of Greg Weiss.

FIGURE 4.18 The exposed straw on the second floor, not yet coated with stucco, contrasts with the finished lower level. Courtesy of Greg Weiss.

Nevertheless, straw—the waste product of grain production—exists largely because industrial machines cleared forests and natural prairies to plant fields of nonindigenous crops that survive with the help of petroleum-based fertilizers, herbicides, and pesticides. The embodied energy of creating and operating farm machinery and the energy associated with the production of fertilizers and other additives are not generally counted against straw bales because that energy is expended to create the grain, not the straw, which may or may not be used. Although straw is considered to have a low embodied energy, its availability as a building material exists because of energy-intensive agricultural practices.

CASE STUDY 4.2

Straw-Bale Timber Frame House

Location: Wisconsin (46° N latitude)

Year built: 2004

Size: 800 ft²

Cost: $38,500 ($48/ft²)

Description: This two-story, two-bedroom straw-bale home is part of a 60-acre organic farm and homestead. Situated in a small grove of trees, the house rests on a concrete slab foundation that serves as its finished floor. Two inches of extruded polystyrene foam insulate the perimeter of the foundation. However, the 5-inch-thick slab itself is uninsulated and remains cool throughout the summer, occasionally resulting in condensation during humid weather.

A timber frame structure made of locally milled pine supports a conventionally framed steel roof. Locally grown straw bales fill around the timber frame and form the walls, with floating 2- × 6-inch stud frames set into the straw for windows and doors. Painted stucco over 1-inch hex wire mesh seals the straw and finishes the interior and exterior walls. Two-foot roof eaves direct rainwater away from the walls and reduce problems associated with moisture. The owners report very few external maintenance requirements, limited to occasional caulking of small cracks in the stucco.

Double-hung windows with double-pane low-emissivity glazing allow excellent ventilation control, while providing an economical compromise on efficiency. Lacking compression seals, double-hung windows permit a relatively high level of air infiltration, a welcomed attribute in this case because the

With the roof in place and supported by a timber frame structure, the owners and volunteers assemble straw-bale walls with reduced risk of rain. Courtesy of Autumn Kelley and Chris Duke.

home includes no form of centralized ventilation. A wood stove provides the only source of heat, consuming two cords of wood per year. The owners accommodate hot summer weather by using movable box fans and by opening windows at night and closing them during the day. The siting of the building within a grove of trees prevented the application of passive solar design.

The 24-inch-thick straw bales provide a wall insulation of R-44, and blown-in cellulose offers roughly R-60 in the attic.

The owners and volunteers did all of the work except for pouring the concrete slab, erecting the timber frame structure, and installing the plumbing and electrical systems. This allowed personal touches and significantly reduced the cost of construction. However, it entailed a great deal of labor, occupying most of a year.

Straw-bale walls provide ample insulation to weather the cold winters of northern Wisconsin. Courtesy of Autumn Kelley and Chris Duke.

Chapter Summary

Alternative construction is an important aspect of green building. Utilizing natural and recycled materials, alternative construction has proven to be practical, economical, and comfortable. Local resources minimize the embodied energy and reduce harmful impacts on the environment, and most natural materials are free of the volatile organic compounds that frequently plague conventionally built homes.

Although earthen materials typically lack the same level of insulation as modern products, thick walls and large thermal mass allow earthen structures to be comfortable in many climates. The durability of alternative construction is well established, with structures still in use after hundreds of years.

For many, however, it is the freedom from standardized materials, the ability to create something by hand, and the chance to custom design spaces that make alternative construction appealing.

Table 4.1 summarizes the advantages and disadvantages of the different forms of alternative construction discussed in the text.

Table 4.2 compares the standard insulation values of conventional construction to the insulation values of different forms of alternative construction for finished walls.

Finally, Table 4.3 compares the embodied energy of a hypothetical, alternatively constructed wall panel measuring 4 feet by 8 feet by 18 inches thick to a similar-sized, conventionally built, 6-inch-thick wall panel. Calculations for embodied energy use Table 2.1 and make the following assumptions:

- All construction methods include the embodied energy of transportation typical for the products. Materials gathered on-site will reduce this energy; those shipped farther than typical will increase this energy.
- The high estimates for adobe, cob, and earthbag construction assume machine digging, mixing, and transporting of material. The low estimates assume all material is moved and handled manually with no assistance from energy-using devices.
- Straw-bale construction includes energy to operate the equipment that harvests, bales, and

TABLE 4.1 Summary of alternative construction methods.

CONSTRUCTION METHOD	ADVANTAGES	DISADVANTAGES	DISTINGUISHING FEATURES
Adobe	• Very high thermal mass • Fireproof • Nontoxic • Very low embodied energy • Durable	• Poor insulation value (R = 0.2 – 0.5 per inch) • Labor intensive • Poor earthquake performance • Performs best in arid climates where bricks dry quickly • Must be protected from moisture	• Thick walls and deep window sills • Well adapted for climates with large daily temperature swings
Cob	• Very high thermal mass • Fireproof • Nontoxic • Very low embodied energy • Durable • Resistant to cracking	• Poor insulation value (R = 0.3 – 1.0 per inch) • Labor intensive • Must be protected from moisture	• Thick, rounded and curving walls • Permits varied textures and forms • Sculpture can be integrated into walls
Cordwood	• Moderate thermal mass • Uses scrap wood • Nontoxic • Low embodied energy	• Transverse log arrangement reduces insulation value • Labor intensive • Limited Trombe wall effect due to air space • Cracking and gapping may occur around logs	• Aesthetics of exposed log ends • Can be built by an individual
Earthbag	• Very high thermal mass • Can utilize any inorganic soil • Stands up to severe weather and earthquakes • Fireproof • Serves as own foundation	• Labor intensive • Limited windows and doors • Polypropylene sacks must be protected from sunlight	• Dome-shaped roof
Rammed-earth tire	• Very high thermal mass • Reuses discarded materials • Fireproof • Serves as own foundation	• Extremely labor intensive • Heavy equipment and trucks required • Limited in design	• Partially subterranean • Made of discarded automobile tires • Passive solar design • Built-in bays
Straw bale	• Moderate thermal mass • Uses agricultural waste product • Nontoxic • Very low embodied energy • Durable if kept dry	• Moderate labor required • Must be protected from moisture • Straw is available because of energy-intensive agricultural practices	• Thick walls • Flexible design options

© 2016 Cengage Learning®.

TABLE 4.2   R-value comparison.

CONSTRUCTION METHOD	R-VALUE (PER INCH)	TYPICAL WALL THICKNESS (INCHES)	TOTAL R-VALUE FOR TYPICAL WALL
Conventional Construction			
Fiberglass insulation	3.1	3.5–5.5	11–17
Cellulose insulation	3.7	3.5–5.5	13–20
Alternative Construction			
Adobe	0.2–0.5	10–30	2–15
Cob	0.3–1.0	10–30	3–30
Cordwood	1.0–1.5	12–24	12–36
Earthbag	0.2–1.5	8–16	2–24
Straw bale	1.8–2.0	15–24	27–48

© 2016 Cengage Learning®.

TABLE 4.3   Estimated embodied energy (without internal or exterior finish surfaces).

CONSTRUCTION METHOD	EMBODIED ENERGY (1000 BTU)	R-VALUE
Adobe	0–1000	4–9
Cob	0–800	5–18
Cordwood	730	18–27
Earthbag	180–1200	2–24
Straw bale	100	36
Conventional 6″-thick wall (2″ x 6″ studs with cellulose insulation)	290	20

© 2016 Cengage Learning®.

ships straw to the site. It also includes polypropylene twine and 1-inch hex wire netting to bind bales together.

- Cordwood construction assumes waste wood is gathered locally, transported by vehicle, and cut by chain saw. It also assumes the use of cement mortar and cellulose insulation.

- Conventional construction assumes a 6-inch-thick wall with 2- × 6-inch studs every 16 inches, a double top plate and single bottom plate, ½-inch oriented strand board (OSB) sheathing, 8d common nails spaced every 6 inches in the sheathing, and cellulose insulation between studs.

Review Questions

1. In what climatic regions is each of the following techniques best suited for residential home construction? Explain your reasoning.
   a. Adobe and cob
   b. Cordwood
   c. Earthbag
   d. Earthship
   e. Straw bale

2. Many natural products can be baled like straw and used for building. These include native grasses, leaves, and shredded bark. What are the advantages of using nonstraw bales? What are the disadvantages?

3. Traditional earthen structures had much thicker walls than is common with conventional construction today. State several reasons for this.

4. Many forms of alternative construction require coatings on the interior and exterior walls for protection. Natural plasters can be made from sand mixed with clay, lime, or gypsum and sometimes fibrous materials like cut straw or manure. Natural plasters can often be dug and mixed at the building site, giving them low embodied energy. However, these finishes require substantially more yearly maintenance than stucco, which is made from Portland cement and sand and has a much higher embodied energy. Discuss the benefits and drawbacks of natural plasters versus stucco. Which is better in the long run?

5. Some builders of rammed-earth tire structures use aluminum cans imbedded in concrete to form non-load-bearing walls. The aluminum cans reduce the amount of concrete required to build the wall and remove leftover cans from the waste stream. However, aluminum cans represent one of the easiest and most economical materials to recycle. Compose an argument for or against their use, justifying your position.

6. Some proponents claim Earthships are appropriate for any climate. Use the six green building principles to assess the benefits and limitations of Earthship design for the four climates indicated in Figure 1.16.

7. Describe a situation in which cob construction would be superior to cordwood construction. In what situation would cordwood be superior to cob?

8. Alternative construction originated from the need to build shelters with materials at hand without the advantages of conventional materials now available. Select one of the six forms of construction discussed in this chapter and discuss ways in which it can be improved using conventional materials. If it cannot be improved, explain why.

9. What style of construction, including but not limited to those discussed in this chapter, would you recommend for building a house in hot, humid climates. Why?

10. Consider three people living outside of Pendleton, Oregon, an arid prairie where the annual average temperature and precipitation are 52°F and 13 inches, respectively.

Each wishes to build a green home. Select an appropriate technique for each person, and justify your reasoning.
   a. An unemployed handyman who has very little money but plenty of time
   b. An individual who has only a modest income and very little free time
   c. A wealthy individual who has no time to contribute to the project but is willing to spend what is necessary

11. Individual preferences greatly influence priorities placed on each of the six green building principles. For someone who wishes to feel connected to the outdoors through abundant windows on multiple sides, what potential compromises are required for a cold climate home? How would you best accommodate this preference?

12. In what climate is a great deal of thermal mass generally not desirable? What form of construction would you recommend for this climate?

13. A builder in Montana wishes to dig a root cellar measuring 8 feet wide by 16 feet long into a north-facing embankment. To stabilize the soil and provide internal structure, yet provide adequate moisture to store food, what building technique(s) would you recommend? Why?

14. What type of construction would be appropriate for a sauna built in northern Minnesota, where the average mean temperature is 37°F? To be useful at relatively short notice, saunas must heat up quickly.

15. Greenhouses can extend the growing season considerably in cold climates by keeping seedlings from freezing. They work best when maintaining relatively constant temperatures. What building technique(s) would you recommend for a cold and wet climate? Explain your reasoning.

16. Companies that manufacture engineered wood products use logs that are unsuitable for sawmills to create a vast array of timber products, including OSB and laminated lumber. In the past, loggers left unsuitable wood to rot, taking only high-quality logs. The abundance of scrap wood helped establish cordwood construction as a building technique, and cordwood

construction remains popular today despite technology that can utilize this scrap wood. The quantity of wood used in a single cord-wood structure could be manufactured into engineered products to provide lumber for three to five similarly sized structures using conventional techniques. Using the principles of green building, assess the merits and limitations of cordwood construction.

17. As society becomes equipped to recycle more and more of its valuable resources (such as automobile tires, aluminum cans, and pulp-wood), how does this affect our consideration of green building materials?

18. In an effort to promote sustainability, suppose the governor of Los Angeles decides that all new structures within the city must be made from adobe. What consequence would this have for the building industry and the environment?

19. If a single form of alternative construction were to be used in all new construction across the United States, what form, if any, would you recommend? Explain your reasoning.

20. What would be required for alternative construction to go mainstream? Should it go mainstream? Why or why not?

Practice Problems

1. A 1400-square-foot rammed-earth tire structure requires about 1000 tires. The owner plans on driving a pickup truck to the local dump to acquire the tires free of charge. Assume the truck gets 15 miles per gallon and is capable of hauling 20 tires at a time.
   a. If a round-trip to the dump is a 30 miles, how many gallons of gas will this require?
   b. At the current price for gasoline, what will this cost the owner?
   c. For two people working together, suppose it takes an average of 20 minutes to position and fill each tire. How many of hours of sweat equity will it take to position the 1000 tires?
   d. If the owner instead hires contractors at $60/hour for two workers, what will the labor of erecting the walls cost?

2. The energy required to create a typical automobile tire is roughly 48,000 Btu/lb. The energy retrieved by burning or gasifying rubber is only about 18,000 Btu/lb, less than half the embodied energy of the tire. However, recycling rubber and using it as feedstock in new products returns as much as 30,000 Btu/lb. Consider a rammed-earth tire structure that uses 1000 tires, each of which weighs 22 pounds.
   a. Instead of being buried, if these tires were recycled and used as feedstock for new

products, what quantity of energy could be saved?
   b. The embodied energy of polypropylene feed sacks used in earthbag construction is roughly 4800 Btu/bag. If it takes 1200 bags to create the same structure as 1000 tires, how much energy (in Btu) would be saved by recycling the tires and purchasing polypropylene bags? Assume barbed wire adds an additional embodied energy of 190,000 Btu to the structure.

3. Consider an 18-inch-thick cordwood wall measuring 20 feet long and 8 feet high that is part of a heated building. Also consider a conventionally built wall of the same length that is insulated with 5½ inches of densely packed cellulose.
   a. Using an R-value of 1.2/inch for cordwood construction, determine the heat loss (Btu/hr) for a temperature difference of 25°F across the wall.
   b. Estimate the average R-value of the conventional wall and its heat loss with similar conditions.
   c. The average annual temperature in Pendleton, Oregon, is roughly 25°F cooler than the 72°F internal temperature most people enjoy. Use this to estimate the total annual heat loss through the two wall types.

d. If every 100,000 Btu of energy cost $2.20, what is the annual cost for each of the two walls?

e. If the conventionally built cellulose-insulated wall is 18 inches thick rather than 5½ inches thick, what would be its R-value?

4. The embodied energy of cordwood construction depends on many things. Consider a cordwood wall measuring 20 feet long and 8 feet high that consists of local scrap wood. Assume the owner debarked and positioned the logs by hand and mixed cement mortar out of a wheelbarrow. Assume 13 gallons of gasoline were required to collect and haul the logs out of the woods and cut the logs to length. Gasoline has an embodied energy of roughly 115,000 Btu/gal. Assume 35 cubic feet of cement mortar bind the logs together and 45 cubic feet of cellulose insulation fill the center cavity.

a. Determine the total embodied energy of the cordwood wall.

b. Find the embodied energy for a similar conventionally built wall. Assume the wall contains 11 2- × 6-inch studs, each 8 feet long; 5 sheets of ½-inch-thick 4- × 8-foot OSB; 5 sheets of ½-inch-thick 4- × 8-foot drywall; and 62 cubic feet of cellulose insulation. Use Table 2.1.

Energy

Source: David Parsons/NREL.

Introduction

FIGURE 5.1 Wood heat is one of the simplest and oldest forms of energy utilized by people. This traditional wood stove provides warmth to a home in northern Wisconsin. Courtesy of Scott Grinnell.

Energy and its use determine the shape of human civilization. Energy enables the manufacture and transport of goods and materials necessary for all aspects of society—from food, clothing, and building products to modern machinery and technology. The form of energy used to fulfill humanity's demands greatly affects the health of society and the environment.

The first forms of energy people made use of—beyond their own muscles and those of animals—included firewood and dried dung for cooking and heating, falling water to grind grain and cut lumber, and wind to run pumps and turn mills (see Figure 5.1 through Figure 5.3). The shift to fossil fuels in the eighteenth century set the stage for the Industrial Revolution. People discovered that fossil fuels contained far greater energy density than traditional sources. Coal, in particular, powered steam engines, fueled factories, fired blast furnaces, and greatly advanced industrial opportunities. Traditional energy sources waned as the potency of fossil fuels became established, and today coal, oil,

FIGURE 5.2 Thousands of water wheels once dotted the countryside, most grinding grain into flour. This one is located near Las Vegas, New Mexico. Courtesy of Scott Grinnell.

FIGURE 5.3 For hundreds of years, people have used windmills to pump water and grind grain. This traditional mill is in the Netherlands. © meirion matthias/Shutterstock.com.

and natural gas satisfy most of the world's energy demands. Interestingly, all of this energy—whether derived from wood, water, wind, or fossil fuels—comes from a single ultimate source: sunlight.

Solar Energy

The sun delivers more energy to the earth in an hour than all of humanity consumes in a year. The land, ocean, and atmosphere absorb most of this energy, but clouds, ice fields, and other reflective surfaces return roughly 30% back into space (see Figure 5.4). Solar energy is the foundational resource that drives the planet's complex systems and manifests itself in a variety of forms:

- **Biomass energy**: Derived from living organisms, biomass includes timber and agricultural products, livestock manure, and municipal garbage. Through

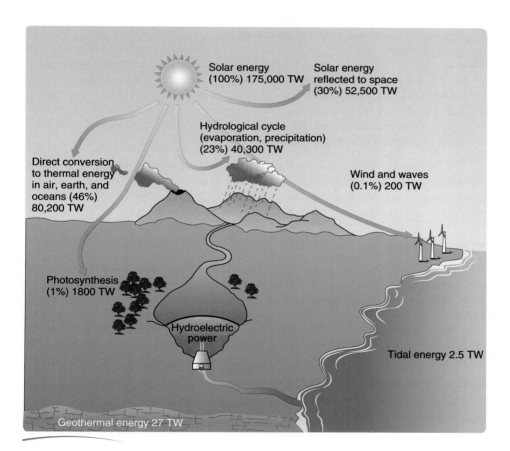

FIGURE 5.4 Sunlight provides nearly all of the energy available on earth. Some of the sun's energy goes into evaporating water to drive the hydrological cycle; some goes into creating wind, waves, and ocean currents; and some provides energy for photosynthetic plants. 1 TW = 1 trillion watts (1×10^{12} watts). © 2016 Cengage Learning®.

photosynthesis, plants absorb sunlight and carbon dioxide to create the carbohydrates that form the building blocks of wood and other organic material. The sun is ultimately the source of all biomass energy.

- **Hydro energy**: As sunlight warms the oceans and continents, some of its energy evaporates water and drives the hydrological cycle. Precipitation falling in mountains and high-altitude plains gathers in rivers and rushes back to lakes and oceans. This flow can rotate wheels to grind grain or spin turbines to produce electricity. Sunlight, therefore, is the source of all hydro energy.
- **Wind energy and ocean current energy**: The sun's more vertical illumination of the tropics and oblique illumination of the poles result in uneven heating of the planet. The differences in temperature set into motion enormous currents of air and water that give rise to wind and ocean currents. Furthermore, wind blowing across open bodies of water can generate waves. Hence, the sun is responsible for all wind, wave, and ocean current energy.
- **Fossil fuels**: Fossil fuels represent millions of years of concentrated solar energy. Prehistoric algae, bacteria, and plants absorbed sunlight, stored energy through photosynthesis, and were buried and compacted under layers of sediment. This concentration of energy makes fossil fuels more potent than renewable forms of energy. It also makes the use of fossil fuels more problematic. The burning of fossil fuels since the Industrial Revolution—a span of a couple hundred years—has released millions of years of stored carbon into the atmosphere, increasing atmospheric carbon dioxide concentrations and presenting serious global challenges. Fossil fuels, therefore, represent prehistoric solar energy.

Sunlight, in its direct and derived forms, accounts for roughly 99.97% of all energy available on earth. Without sunlight, there would be no rain or wind or running water. There would be no soil, no plants, and no life as we know it. Nevertheless, two nonsolar sources of energy remain important: tidal and geothermal.

Tidal Energy

Tidal energy results from the gravitational forces acting on the earth from the moon and sun. These forces pull on the earth and its oceans, raising a tidal bulge. As the earth spins on its axis, the tidal bulge remains nearly stationary and drags against the continents. The movement of the oceans relative to the continents gives rise to the daily tides and provides another source of available energy. In many coastal regions, tidal energy can be substantial (Figure 5.5).

Geothermal Energy

Resulting primarily from the decay of radioactive elements within the earth, **geothermal energy** drives the tectonic motions of the continents, generates volcanoes and earthquakes, and occasionally penetrates close enough to the surface to heat groundwater and create geysers and hot springs (see Figure 5.6 and Figure 5.7). In some locations, heated water can drive turbines and generate electricity. All together, geothermal energy provides roughly 0.016% of the earth's total available energy at the surface, though only a small fraction of this could ever be harnessed for energy production.

FIGURE 5.5 The Bay of Fundy, between the Canadian provinces of New Brunswick and Nova Scotia, experiences the largest tides in the world, varying by more than 50 feet. This change in water level offers a source of energy for many coastal communities around the world. **(a)** © GVictoria/Shutterstock.com. **(b)** © Melissa King/Shutterstock.com.

FIGURE 5.6 Geysers sometimes form when geothermal energy heats water seeping into fissures. This geyser is located in Yellowstone National Park. Courtesy of Chris Grinnell/ Bearing the Light LLC.

FIGURE 5.7 Geothermal energy heats plumes of molten rock that can erupt at the earth's surface as volcanoes. © beboy/Shutterstock.com.

Together these three sources—solar, tidal, and geothermal—provide essentially inexhaustible and continuously renewable energy. Each source has advantages and shortcomings, and different forms are limited to particular locations, seasons, or times of day. No single source of renewable energy promises to replace current fossil fuel consumption. Satisfying future energy demands will require a combination of sources along with increased efficiency and conservation.

This chapter introduces the fundamental types of energy, discusses the principles that govern energy use, and then examines two of the most relevant aspects associated with energy: calculating electrical energy usage and assessing the available solar resource.

Types Of Energy

The three fundamental sources of energy—solar, tidal, and geothermal—represent opportunities for people to harness energy and make it do useful work. These "sources" do not *create* energy, and using energy does not *destroy* it. A fundamental rule of nature, called the **First Law of Thermodynamics**, states:

> *Energy is neither created nor destroyed. It is simply converted from one form to another.*

The act of using energy is a process of conversion, not consumption. The total energy of the universe remains constant—the same today as it has always been. While it is common to speak of *consuming* energy, it simply means converting energy from one form to another.

Energy can exist in four fundamental types: kinetic, gravitational, electrical, and nuclear.

Kinetic Energy

Kinetic energy is the energy of motion. All moving objects—a motorcycle racing down the road (Figure 5.8), water rushing over a dam, wind sweeping across a prairie, and the rotating shaft of a turbine—possess kinetic energy. The faster something moves and the larger, more massive it is, the greater its kinetic energy.

FIGURE 5.8 Kinetic energy is the energy of motion. © okx/Shutterstock.com.

In addition to the motions of observable objects, another less obvious form of kinetic energy takes place on the molecular level. Thermal energy—the transfer of which we feel as heat—is the average random kinetic energy of atoms and molecules that make up matter. Temperature measures the average molecular kinetic energy: the hotter the material, the greater its molecular kinetic energy (see Figure 5.9). This form of energy is chaotic and highly unorganized.

Gravitational Potential Energy

Gravitational potential energy is the energy associated with the position of objects within the presence of gravity (see Figure 5.10). On earth, the higher something is lifted and the greater the object's mass, the larger its gravitational potential energy. Lifting up bales of straw requires more effort than lowering them. Similarly, a ball at the top of a steep hill has gravitational potential energy that converts into kinetic energy as it rolls down.

Electrical Energy

Electrical energy is the energy associated with the positions and motions of charged particles. Every atom consists of a cloud of electrons surrounding a central charged nucleus. Electrical energy manifests itself in three common forms: electricity, chemical energy, and electromagnetic energy.

- Electricity is the organized motion of free electrons. Materials such as metals allow electrons to move freely from atom to atom and commonly serve to convey electrons in many modern applications. From computers and cell phones to motors and lightbulbs, electricity serves as the basic energy source for much of technology. Lightning is a natural form of electricity (Figure 5.11)—the movement of electrons through the earth's atmosphere.
- Chemical energy includes the stored energy associated with food, gasoline, matches, and batteries. The location of an electron and its state within an atom or molecule determine its chemical energy. Whenever atoms or molecules establish or break bonds or alter their arrangement with each other, the locations and distribution of electrons change. The release of chemical energy is the change in electrical energy associated with the rearrangement of electrons. For example, striking a match, as in Figure 5.12, causes a rapid repositioning of electrons as atoms of sulfur and phosphorous bond with oxygen.

FIGURE 5.9 Thermal energy is the random kinetic energy of atoms and molecules. The iron atoms composing the tip of this red-hot blade are moving more rapidly on average than atoms in the dull handle. Courtesy of Scott Grinnell.

FIGURE 5.10 Gravitational potential energy exists whenever an object is able to fall under the influence of gravity. The difference in height between the canyon floor and the tightrope endows the man with gravitational potential energy. © istockphoto.com/Michael Svoboda.

FIGURE 5.11 Electricity is the coherent motion of electrons, whether created naturally in the atmosphere or generated within power plants. On this night, both illuminate the city. © Jhaz Photography/Shutterstock.com.

FIGURE 5.12 The energy released with the burning of a match comes from electrical energy stored as chemical bonds between atoms. Courtesy of Scott Grinnell.

- **Electromagnetic energy** includes visible light, ultraviolet radiation, infrared radiation (heat), microwaves, radio waves, and X-rays. When an electron accelerates, its changing electric field creates a magnetic field. The changing magnetic field in turn generates another electric field, and the two fields—each recreating the other—produces an electromagnetic wave that can propagate indefinitely. Electromagnetic waves are unique in their ability to pass through the emptiness of space, requiring no material medium to maintain the wave (Figure 5.13).

Nuclear Energy

Nuclear energy is the energy associated with the number of protons and neutrons bound together in the nucleus of an atom. Although nuclear energy is rarely experienced in everyday life, the fusion of hydrogen into helium within the sun's core—converting nuclear energy into electromagnetic energy—is the process that creates the sunlight necessary for all life on earth. The reverse process, the splitting of large elements like uranium into smaller ones, is the foundation of existing nuclear fission power plants and atomic bombs (see Figure 5.14).

FIGURE 5.13 Sunlight is a visible form of electromagnetic energy. Most forms, like the cell phone signals relayed by this tower, go undetected by the human eye. Courtesy of Scott Grinnell.

FIGURE 5.14 The detonation of an atomic bomb releases a form of energy rarely experienced on earth: the energy binding together the atomic nucleus. © Sergey Nivens/Shutterstock.com.

Energy Conversion

Energy can be converted from one form into any other form. Consider the following transformations:

- Photosynthesis in a leaf converts electromagnetic energy into chemical energy when utilizing sunlight to make carbohydrates.
- Burning a piece of wood converts stored chemical energy into thermal and electromagnetic energy by providing heat and light.
- The motor of a ceiling fan converts electrical energy into kinetic energy by moving air with rotating blades.

- The whirling impeller of a pump converts kinetic energy into gravitational potential energy by lifting water into an elevated tank.
- A solar cell converts the electromagnetic energy of sunlight into electricity.

Although energy *can* be converted from any form into any other, it will *tend* to move to a form that is less organized than the original. Thermal energy, the chaotic kinetic energy of randomly moving molecules, is the least organized form of energy and is the natural outcome of energy conversions. This tendency toward disorder is another fundamental rule of nature, called the Second Law of Thermodynamics:

> *In any conversion, the energy of a system always changes from a more ordered form to a less ordered form.*

The unordered state of thermal energy limits its ability to convert into other forms and, therefore, reduces its overall usefulness. For this reason, thermal energy is considered low-grade energy. Conversely, electricity (the coherent motion of electrons) is much more highly organized and its applications much more diverse. Electricity easily converts into other forms and is, therefore, considered high-grade energy.

A low-grade energy like thermal energy can be converted into high-grade energy like electricity but always with a penalty of creating more disorder elsewhere. Although some of the thermal energy becomes more ordered, an even larger portion of it becomes increasingly disordered, resulting in more overall disorder (see Figure 5.15). This becomes clear when considering a conventional coal-fired power plant in the next section.

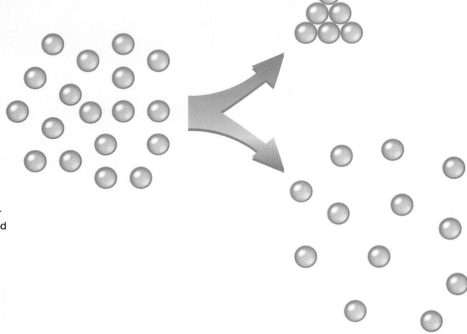

FIGURE 5.15 Whenever some energy is converted into a more ordered form, an even greater proportion becomes increasingly disordered. Every process of energy conversion creates greater overall disorder.
© 2016 Cengage Learning®.

Conventional Power Generation

The Second Law of Thermodynamics sets inherent limits on the amount of electricity or useful work that can be extracted from a heat source. No clever design or marvelous feat of engineering can circumvent this inherent limit. It is fundamental to nature and unalterable.

The burning of coal in a conventional power plant converts its chemical energy (stored from millions of years of photosynthesis) into heat. This heat boils water and produces steam under high pressure. Steam molecules collide with specially shaped turbine blades and transfer some of their kinetic energy to the turbine. The rotating turbine spins the shaft of a generator, and magnetic fields within the generator convert the shaft's kinetic energy into electricity.

While some of the thermal energy of steam becomes more ordered as the turbine converts it into electricity, an even larger portion of the thermal energy becomes even more disorganized in the form of waste heat. Most conventional power plants simply dump this waste heat into nearby lakes or rivers, as shown in Figure 5.16.

An illustrative example of this process, shown in Figure 5.17, provides typical efficiencies for each process of energy conversion in a conventional coal-fired power plant. The **efficiency** represents the ratio of the *output* energy to the *input* energy and is, therefore, always less than 100%.

Machines grind coal into powder and blow it into a combustion chamber with a controlled mixture of air. As the coal ignites and its chemical energy converts to thermal energy, some of the heat necessarily escapes up the smokestack to carry away flue gases (shown as a 14% loss, meaning 86% of the thermal energy is still available to boil water). Imperfections in insulation, friction within moving parts, and electrical resistance within wires all add inefficiencies to the process and contribute to waste heat (approximately 6% lost). The greatest inefficiency, however, occurs where the turbine extracts the steam's thermal energy to create kinetic energy of the rotating shaft (a loss of 56%). Only a fraction of the low-grade thermal energy can be converted to higher-grade kinetic energy; no matter how perfect the engineering, the process is limited by the Second Law of Thermodynamics. The remainder of the thermal energy becomes waste heat that warms rivers or lakes or (in more resourceful designs) flows through underground pipes to warm city buildings. The dispersion of the thermal energy as waste heat makes it even more disordered than the original thermal energy of the steam, resulting in more overall disorder despite creating some electricity.

FIGURE 5.16 All conventional power plants, whether fueled by the combustion of coal or wood chips, convert thermal energy into electricity by rotating a turbine connected to the shaft of a generator. Courtesy of Scott Grinnell.

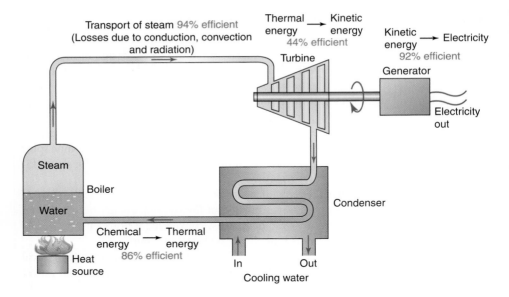

Overall efficiency: 86% × 94% × 44% × 92% = 33%

FIGURE 5.17 The overall efficiency of converting the stored chemical energy of coal into electricity is roughly 33%. Waste heat accounts for the remaining 67% of the original chemical energy. The greatest inefficiency occurs during the conversion of thermal energy into kinetic energy, limited by the Second Law of Thermodynamics. © 2016 Cengage Learning®.

Finally, the conversion of kinetic energy into electrical energy in the generator is about 92% efficient, losing an additional 8% as waste heat.

The total efficiency of the entire process is the product of the efficiencies at each transition, or 0.86 × 0.94 × 0.44 × 0.92 = 0.33. Of the stored chemical energy present in coal, only 33% is ultimately converted into electricity. The remaining 67%—by far the majority—is sacrificed as waste heat. Transmission losses between the power plant and the point of use further decrease this efficiency.

The principle of generating electricity from high-temperature thermal energy is the same for all conventional power plants, whether the fuel source is coal, wood chips, natural gas, or enriched uranium (in a nuclear power plant). In addition, the same principle applies to the internal combustion engine of automobiles, though the kinetic energy delivered to an automobile tends to be only about 20% of the chemical energy stored in the gasoline. The remainder is shed to the environment as waste heat.

Renewable Energy

Renewable energy differs from fossil fuel energy in that it is replenished at least as quickly as it is consumed. The rate at which sunlight reaches the earth's surface is ten thousand times greater than humanity's combined consumption, and harnessing it does not diminish its supply. Conversely, fossil fuels are nonrenewable because the rate of ongoing fossil fuel production within the earth is negligible compared to humanity's consumption of it.

Renewable energy currently provides only 9% of total U.S. energy demands, dominated by hydroelectric, wood, and biofuels (see Figure 5.18). In recent years, wind energy has grown the fastest and promises to provide inexpensive energy that competes in price with fossil fuels. Ongoing developments in solar electric technology and biomass fuel production also promise greater future contributions.

The dominant forms of renewable energy include *solar hot water, photovoltaic (solar electric), wind, hydroelectric*, and *biomass*. Subsequent chapters address each of these forms in greater detail.

Solar hot water systems convert the electromagnetic energy of sunlight into thermal energy. This process is potentially the most efficient of all renewable energy systems (achieving a theoretical efficiency of 100%) because high-grade electromagnetic energy readily converts into low-grade thermal energy. The actual efficiency of solar hot water systems is substantially less than this theoretical limit and varies dramatically with design and location, but it is often between 60% and 70%. The energy not converted into usable hot water is reflected from the collectors, reradiated back to the environment, or lost by conduction and convection.

Photovoltaic systems convert the electromagnetic energy of sunlight directly into electricity. Most available solar electric systems offer a maximum conversion efficiency of roughly 20%, but this value is limited more by engineering than the laws of

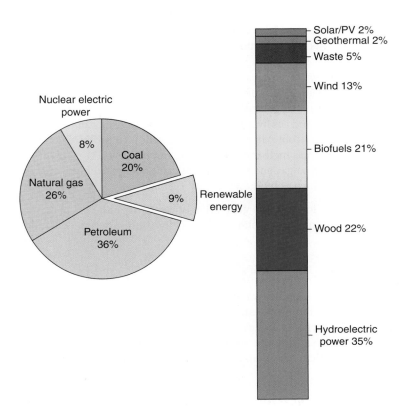

FIGURE 5.18 Renewable energy currently supplies a relatively small proportion of the United States' total energy budget. However, declining reserves of fossil fuels and concerns over climate change present considerable opportunities for future growth in renewable energy. © 2016 Cengage Learning®. *Source:* Adapted from U.S. Department of Energy, U.S. Energy Administration.

thermodynamics. It may be possible to attain conversion efficiencies as high as 85% with improved technology.

Wind turbines extract kinetic energy from moving air to produce electricity and are limited by a 59% maximum theoretical efficiency. Air cannot lose all of its kinetic energy, or it would come to a standstill behind the turbine and pile up, blocking subsequent air. Clearly, air must keep moving beyond the turbine, and the amount that does so must retain at least 41% of its original kinetic energy. Modern wind turbines perform at less than this theoretical maximum, obtaining roughly 35%–50% efficiency, depending on wind conditions and turbine design.

Hydroelectric turbines extract both kinetic and gravitational potential energy from falling water to produce electricity, allowing a greater efficiency than wind turbines—as high as 90%. The high efficiency and predictability of hydroelectric power plants made them the first widely distributed renewable energy system.

Biomass energy varies greatly, depending on the resource and the method of conversion. The maximum efficiency of converting sunlight into biomass through photosynthesis is roughly 5%, though most green plants achieve a much lower efficiency, usually less than 1%. Nevertheless, the area covered by photosynthesizing organisms on continents and oceans makes this potential enormous. Different types of biomass utilize different conversion techniques:

- Biomass in the form of wood chips, municipal garbage, or solid agricultural waste is often burned in power plants to create electricity, similar to conventional coal-burning plants. This conversion has roughly the same efficiency as conventional coal-burning power plants, though moisture and contaminants can reduce the efficiency to 20% or less.
- The solid biomass mentioned above can also be heated in the absence of oxygen (which prevents combustion) to produce combustible gas or liquid fuels. These fuels can be used directly for heating or subsequently burned to produce electricity.
- Biomass in the liquid form, such as livestock manure or grain slurry, can be biologically converted through anaerobic digestion or fermentation to produce combustible gas or liquid fuels, such as methane and ethanol, respectively.
- Biomass in the form of microalgae and oilseed crops can undergo chemical conversion to create liquid fuels, such as biodiesel.

Biomass is a promising source for both the generation of electricity and the creation of liquid fuels that may be able to replace gasoline and diesel for transportation demands.

Fundamentals of Electricity

With a few exceptions—such as solar hot water systems and the production of combustible gas and liquid fuels from biomass—energy conversion systems create electricity. As one of the most versatile forms of energy, electricity plays a vital role in sustaining a technological society. Understanding the fundamental concepts of electricity and acquiring the skills to make simple calculations are necessary to assess the performance of common energy systems.

Ohm's Law

Electricity is the movement of electrons. Electrons move when exposed to an electric field, which can be due to neighboring electrons (or other charged particles) or changes in magnetic fields. The pressure compelling an electron to move is called the voltage. The greater the voltage, the faster electrons tend to move. As electrons flow through a material, such as a wire, they collide with atoms in the material and experience resistance. Resistance slows electrons and limits the number of electrons that flow past a point each second, called the current. Ohm's Law summarizes this relationship:

$$Current = Voltage/Resistance$$

Electricity passing through a wire can be compared to water flowing through a pipe. The pressure in the pipe (voltage) establishes the driving force and determines the speed at which water flows, but the size and smoothness of the pipe (resistance) limit the quantity of flow (current) and reduce the pressure (voltage). Materials that allow electrons to pass with relatively little resistance are called conductors. Copper and aluminum are good conductors and are most commonly used to convey electricity. Electrons moving through a conductor compare to water flowing easily through a smooth pipe.

Electrical Power and Energy

Power represents the *rate* of using energy—that is, the amount of energy used over an interval of time. Electrical power depends on both the voltage (pressure) and the current (quantity of flow):

$$Power = Voltage \times Current$$

The total electrical energy used during a process equals power multiplied by time (in hours):

$$Energy = Power \times Time$$

Standard Units

Standardized units (and symbols) for each of these quantities allow numerical calculations and rapid communication of ideas. Most of the world uses the International System of Units (SI or metric system) for all measurements. The United States has adopted the SI system for electricity but still uses the old English system of British thermal units (Btu) for thermal energy calculations.

Standard units for electricity.			
QUANTITY	SYMBOL	UNIT NAME	UNIT ABBREVIATION
Voltage	V	volts	V
Current	I	amps	A
Resistance	R	ohms	Ω
Power	P	watts	W
Energy	E	kilowatt-hour	kWh
Time	T	hours	h

A kilowatt equals 1000 watts and is a measure of power. For conversion, 1 watt = 3.412 Btu/hr. Similarly, 1 kWh = 3412 Btu.

EXAMPLE 5.1

A compact fluorescent lightbulb (CFL) uses 21 watts of power when plugged into a standard 120-volt outlet. Of the 21 watts, only 4 watts are visible light.

a. What current does the bulb draw?
b. What is the electrical resistance of the bulb?
c. What is the efficiency of the bulb in converting electricity into visible light?
d. If the bulb is left running continuously for 1 week, how much energy (in kWh) does it use?
e. If the electric company charges $0.12/kWh of electricity, how much does this usage cost?

Solution:

a. The current equals the power divided by the voltage, or:

$$I = P/V$$

$$I = 21 \text{ watts}/120 \text{ volts}$$

$$I = 0.175 \text{ amps}$$

b. Using Ohm's Law, the resistance equals the voltage divided by the current, or:

$$R = V/I$$

$$R = 120 \text{ volts}/0.175 \text{ amps}$$

$$R = 686 \text{ ohms}$$

c. The efficiency represents the useful output divided by the total input. Since only 4 watts of the total 21 watts are converted into visible light, the bulb has an efficiency of:

$$\text{Efficiency} = 4 \text{ watts}/21 \text{ watts}$$

$$\text{Efficiency} = 19\%$$

d. Energy equals power multiplied by time:

$$E = P \times T$$

$$E = 21 \text{ watts } (7 \text{ days} \times 24 \text{ hours/day})$$

$$E = 3500 \text{ watt-hours} \times 1 \text{ kW}/1000 \text{ W}$$

$$E = 3.5 \text{ kWh}$$

e. The cost for this amount of electricity is given by:

$$\text{Cost} = \text{Energy} \times \text{Price per kWh}$$

$$\text{Cost} = 3.5 \text{ kWh} \times \$0.12/\text{kWh}$$

$$\text{Cost} = \$0.42$$

EXAMPLE 5.2

The heating elements of a toaster oven have a total resistance of 12 ohms and convert electricity into heat with 100% efficiency. Assume the toaster is plugged into a standard 120-volt outlet.

   a. How much power does the toaster oven use?
   b. What current does it draw?
   c. At $0.12/kWh, what does it cost to run the oven continuously for 3 hours?
   d. How much heat (in Btu) does the oven produce during the 3 hours?

Solution:

   a. Since the current is not provided in this example, we must use Ohm's Law to eliminate current from the equation for power. By Ohm's Law, $I = V/R$. Therefore:

$$P = I \times V$$
$$P = V/R \times V$$
$$P = V^2/R$$
$$P = (120 \text{ volts})^2/12 \text{ ohms}$$
$$P = 1200 \text{ watts}$$

   b. The current can be found in two ways: by Ohm's Law or by using the equation for power. Ohm's Law gives:

$$I = V/R$$
$$I = 120 \text{ volts}/12 \text{ ohms}$$
$$I = 10 \text{ amps}$$

Alternatively:

$$I = P/V$$
$$I = 1200 \text{ watts}/120 \text{ volts}$$
$$I = 10 \text{ amps}$$

   c. The cost for this electricity is given by:

$$\text{Cost} = \text{Energy} \times \text{Price per kWh}$$
$$\text{Cost} = \text{Power} \times \text{Time} \times \text{Price per kWh}$$
$$\text{Cost} = (1200 \text{ watts} \times 1 \text{ kWh}/1000 \text{ watts}) \times 3 \text{ hours} \times \$0.12/\text{kWh}$$
$$\text{Cost} = \$0.43$$

   d. The amount of heat produced by the oven equals the electrical energy used, since the conversion is 100% efficient.

$$\text{Heat} = \text{Electrical energy} \times 100\% \times 3412 \text{ Btu/kWh}$$
$$\text{Heat} = \text{Power} \times \text{Time} \times 100\% \times 3412 \text{ Btu/kWh}$$
$$\text{Heat} = 1.2 \text{ kWh} \times 3 \text{ hours} \times 100\% \times 3412 \text{ Btu/kWh}$$
$$\text{Heat} = 12{,}300 \text{ Btu}$$

Alternating and Direct Current

Electricity begins and ends at a power supply. Along the way, machines and appliances convert electricity into useful work, reducing the voltage with each expenditure. Two distinct types of electricity exist.

- **Direct current** (DC) electricity represents a steady, continuous flow of electrons through a circuit, with each electron traveling the entire length of the circuit. Sources of DC electricity include batteries, photovoltaic panels, and lightning.
- **Alternating current** (AC) electricity reverses direction many times a second, causing electrons to oscillate back and forth but not travel any great distance. An oscillating electron forces its neighboring electron to oscillate, which forces its neighbor to oscillate, and so on, propagating energy through the conductor without any of the electrons traveling through the circuit. Commercial power plants supply AC electricity, which is the standard type of household electricity.

Machines and appliances designed for one type of electricity do not operate using the other type. An **inverter** is a device that converts DC electricity into AC electricity, and a **rectifier** converts AC electricity into DC electricity. As with any conversion, some energy is lost as waste heat during the process.

Solar Resource

The amount of solar energy available at any site varies dramatically with location and season. Calculating the electricity generated by a photovoltaic array, the amount of hot water produced by a solar hot water system, or the yield of a particular biomass crop requires knowledge of the available solar energy, called the **solar resource**. The solar resource takes into account latitude, season, and average weather conditions (such as cloud cover). Equatorial deserts tend to have the greatest solar resource, while cloudy polar regions have the least. Figure 5.19 provides a map of the solar resource for the United States.

Solar Altitude

As discussed in Chapter 3, the earth tilts at 23.5° with respect to its orbit around the sun, producing the annual seasons. The earth's tilt causes the sun's altitude to vary by 47° between summer and winter. For nonequatorial regions, the sun's altitude is the highest in the summer and the lowest in the winter (at the equator, the sun is directly overhead at the spring and fall equinoxes).

The sun's altitude determines its intensity: the more directly overhead, the more intense. This is a result of two factors: (1) The sun's rays disperse over a smaller area when directly overhead, as shown in Figure 5.20, and (2) the sun's rays pass through less of the earth's atmosphere, which absorbs and scatters some of the energy.

Figure 5.21 allows an estimation of the sun's altitude for any latitude throughout the year. The solar altitude for any particular latitude is given by 90° – latitude on the spring and fall equinoxes. At any other time of the year, the solar altitude is equal to the equinox value plus the seasonal adjustment angle provided. For example, the solar altitude on

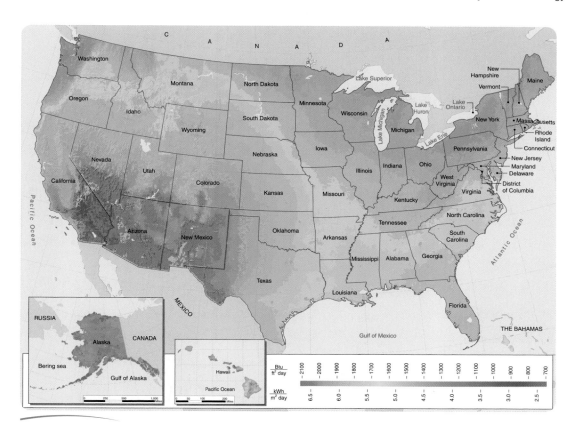

FIGURE 5.19 The solar resource in the United States varies from as high as 2150 Btu/ft^2 in the Southwest to as low as 850 Btu/ft^2 in parts of Alaska. The solar resource takes into account weather conditions (such as cloud cover), the angle of the sun (which depends on latitude), and the relative thickness of the atmosphere. The values shown assume the collector is tilted at the location's latitude. © 2016 Cengage Learning®. *Source:* Adapted from National Renewable Energy Laboratory (NREL).

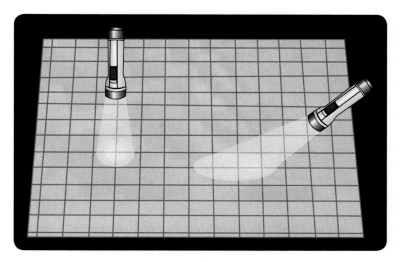

FIGURE 5.20 A beam of light striking a surface obliquely leaves a larger and more diffuse imprint than a more perpendicular beam. Similarly, spreading sunlight over a larger area dilutes its energy and reduces its intensity.
© 2016 Cengage Learning®.

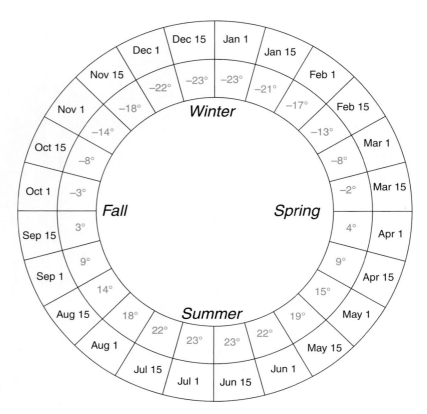

FIGURE 5.21 The solar altitude for any day of the year can be found by adding the seasonal adjustment values (shown in red) to the equinox value of 90° – latitude. The negative numbers between fall and spring reduce the solar altitude.

© 2016 Cengage Learning®.

May 15 at a latitude of 60° N (Juneau, Alaska) is given by 90° − 60° + 19° = 49°. This represents the height of the sun above the horizon at noon on May 15.

The ability of a system to capture the available solar resource depends on its orientation with respect to the sun. Ideally, collectors should face directly toward the sun at all times. This arrangement intercepts the greatest amount of sunlight and minimizes reflection from the collector surface, which becomes more significant at increasingly oblique angles.

Due to the flexible nature of electrical wires, photovoltaic systems have the capability of tracking the sun, maximizing the energy intercepted by the collector. Most solar hot water systems, however, connect rigid pipes to the collectors and cannot move. Hence, the choice of orientation must be made at the time of installation.

For a fixed solar system, whether hot water or photovoltaic, positioning the collector perpendicular to the midday sun maximizes the energy received. This requires adjustment in two directions: (1) orienting the collector to face due south (in the Northern Hemisphere) and (2) tilting the collector to match the sun's altitude. Since the sun's altitude varies by season, solar hot water systems require a choice regarding when hot water is most needed. If hot water is needed equally year-round, then tilting the collector at the location's latitude is ideal. This aims the collector at the sun's altitude on the equinoxes, which is also the annual average altitude. To increase the efficiency in the winter, the collector must be tilted more steeply than the latitude. Conversely, to maximize the summer efficiency, the collector must be tilted less steeply. Since the angle of the sun varies by 23.5° during each season, increasing or decreasing the tilt by approximately half this amount (often rounded to 15°) maximizes seasonal heating. For example, in Juneau, Alaska (60° N latitude), the optimal angle for winter collection is roughly 60° + 15° = 75°. These collector panels are often mounted on the walls of buildings rather than the rooftops.

Shading

An instrument used to determine the degree of shading at a particular site is called a solar pathfinder. This instrument superimposes shadows cast by trees, buildings, and so forth onto a diagram plotting the daily path of the sun for each month of the year. A shadow over any section of the diagram indicates shading will occur at that particular time during that month. Figure 5.22a shows a pathfinder with a diagram appropriate for locations at 43°–49° N latitude. The curved lines plot the path of the sun for each month from 6:00 A.M. to 7:00 P.M. The curves representing the summer months (when the sun is high in the sky) are at the bottom; those representing the winter months (when the sun is low in the sky) are at the top.

Figure 5.22b shows a plastic dome over the diagram in which shadows of nearby trees can be seen. The trees will shade the site in the morning until 8:00–8:30 A.M. during the summer and until around 9:30 A.M. during the rest of the year. Evening shadows begin to block out sunlight around 2:00 P.M. in the winter and about 5:00 P.M. during the summer. The most critical shading, however, occurs during December and January, when the site will be nearly constantly in shadows. This shading will reduce the solar resource during the winter and compromise the effectiveness of solar hot water collectors, photovoltaic systems, and buildings designed for passive solar heating. The shading will be less significant if the trees are deciduous, but even bare branches reduce the sun's intensity.

A solar pathfinder allows accurate measurements of the anticipated solar resource for any site and helps determine the suitability of sites for solar-designed buildings and renewable energy installations.

FIGURE 5.22 (a) Each curved line plots the daily path of the sun for a particular month, with the summer months at the bottom and the winter months at the top. Courtesy of Scott Grinnell.

FIGURE 5.22 (b) Shadows from trees indicate times and dates when shading will reduce the available solar resource. Courtesy of Scott Grinnell.

Chapter Summary

The overwhelming majority of all energy available to people on earth comes from the sun. Sunlight accounts for wind and wave energy, hydro energy, and the energy derived from photosynthetic plants. Fossil fuels represent millions of years of concentrated solar energy, making them energetically denser than renewable forms. Nevertheless, sources of renewable energy offer thousands of times more energy than demanded by all of humanity's activities combined.

Energy is never created, nor is it ever destroyed. "Consuming" energy represents a change from one form to another. During any process of conversion, the total energy of a system always becomes less organized than it was before the conversion. If some energy is made more organized, even more of it will become increasingly less organized. This limits the efficiency at which power plants convert fuel into electricity, which is the orderly motion of electrons. The high degree of order of electricity allows it to convert easily into other forms of energy, giving electricity great versatility and making it essential to our technological society.

Review Questions

1. What form of energy conversion occurs when an ordinary automobile stops at a traffic light? How is this process different for a vehicle with regenerative braking, a type of system integrated into most hybrid vehicles?

2. A child at a park swings back and forth. At what point during the swing is the child's kinetic energy the greatest? When is the child's gravitational potential energy the greatest?

3. Why is it not possible to convert energy from one useful form to another over and over indefinitely?

4. How would living in a house where space and water are heated solely by the sun influence a person's daily living schedule?

5. Suppose a father and son hike to the top of a mountain. If the son weighs half as much as the father, how do their gravitational potential energies compare when standing together at the top?

6. Wood, as a form of biomass, is often considered a renewable energy source. Is this true in all locations? Justify your answer.

7. Ethanol can be made through the distillation of corn. When considering the energy associated with agriculture, transportation, processing, and distillation, the energy expended to produce the ethanol is close to the energy content of the ethanol. Discuss the benefits and drawbacks in using corn to produce ethanol.

8. It is not uncommon for people living in rural India to gather livestock manure, place the manure in appropriately piped cisterns, and generate their own combustible gas through anaerobic digestion. This gas is then used for cooking. Trace the series of energy conversions beginning with sunlight and concluding with the heat given off by the stove.

9. What is the maximum altitude of the sun in Fairbanks, Alaska, on May 5?

Practice Problems

1. The phantom loads of a particular residence include the following: television = 20 W; phone = 8 W; clock = 1 W; computer = 12 W. What percentage of the total monthly electrical usage comes from phantom loads if the house uses 200 kWh per month?

2. For millions of years, plants absorbed sunlight to create biomass. Some of this biomass formed peat bogs that were buried in sediment and compressed into coal. Using the table below, estimate the overall efficiency of producing light with a CFL using electricity generated in a

coal-fired power plant. Begin with the original sunlight. Compare this to south-facing windows in a passive solar home that allow 60% of the visible light to enter and illuminate the building.

Process	Efficiency
Conversion of sunlight to biomass	3%
Burial of biomass in peat bogs	10%
Conversion of peat bogs into coal	4%
Mining of coal	88%
Transportation of coal	92%
Generation of electricity	33%
Transmission of electricity	90%
Conversion of electricity to light	20%

3. What rate of thermal energy (in Btu/hr) is dumped into the environment by a nuclear power plant that consumes 5.0 million Btu/hr of nuclear fuel if the power plant is 35% efficient at generating electricity?

4. Suppose 5 CFLs, each drawing 23 watts, are accidently left on for a month while the owner is away. Assume the generation and transmission of the electricity is 30% efficient. 1 watt = 3.412 Btu/hr.

   a. How much chemical energy from coal did the power plant harness to produce the electricity required to illuminate these bulbs for a month (in Btu)?

   b. How much energy (in Btu) was dumped as waste heat at the power plant?

   c. What volume of cellulose insulation (in cubic feet) has an embodied energy equal to the waste heat (see Table 2.1)?

5. An electric vehicle with a 100-mile range obtains 4.1 miles per kWh of electrical energy, provided by a lithium-ion battery.

   a. How much chemical energy (in kWh) must the battery be capable of storing to permit the vehicle to travel the full 100-mile range?

   b. If the battery is charged with electricity that costs $0.12/kWh, how much does it cost to travel 100 miles?

   c. How does this compare to the cost of a 100-mile journey in a conventional gasoline-powered vehicle that obtains 33 miles per gallon?

6. The charging base for a cordless telephone plugged into a standard 120-volt outlet consumes 12 watts of electricity.

   a. What current does the charging base draw (in amps)?

   b. What resistance does the charging base have (in ohms)?

   c. How much energy is used in a year (in kWh)?

   d. If electricity costs $0.12/kWh, how much does it cost to keep the phone plugged in for a year?

7. Even when turned off, televisions draw electricity. This continuous use of electricity, called a phantom load, allows immediate response when activated by a remote. Suppose a television has a phantom load of 18 watts.

   a. How much energy is consumed in a year due to a phantom load if the television is off an average of 20 hours per day (in kWh)?

   b. At $0.12/kWh, what does this cost for a year?

   c. How much waste heat is produced at a coal-fired power plant to provide electricity for the phantom load if the generation and transmission of electricity is 30% efficient (in Btu)?

8. A refrigerator is a form of heat pump. Electricity runs a compressor that pumps heat out of the refrigerator's interior and dumps it into the surrounding room. The electricity required to run the compressor also becomes waste heat and adds to the heat dumped into the room. Assume a particular refrigerator operates on 120 volts and draws 1.5 amps when running.

   a. How much power does the refrigerator consume (in watts)?

   b. Suppose the refrigerator door is left open, allowing cold air to enter the room. What is the net heating of the room (in Btu/hr)?

   c. Explain why leaving the refrigerator door open won't cool the room.

Solar Hot Water

Courtesy of Scott Grinnell.

Introduction

People have used the sun to heat water since the earliest civilizations. Ancient Egyptians filled earthen pots with water and set them in the sunshine to prepare the water for bathing and cooking. Romans used passive solar design to heat public bathhouses. Early farm laborers filled dark tanks and used the sun-warmed water to soothe tired muscles at the day's end.

While employing the sun's energy to heat water has been part of human society since antiquity, the first modern solar hot water systems did not appear until the 1800s, when running water became common in buildings. These early collectors were simple black-painted water tanks mounted on roofs that heated directly in the sunlight. Even on sunny days, however, these tanks usually did not get hot until afternoon, and lacking any form of insulation, they cooled quickly during the night. In 1891, the first commercial solar water heater, sold under the name *Climax*, improved this model by enclosing the tank in an insulated box sealed behind a pane of glass (see the original advertisement in Figure 6.1). The box reduced heat loss at night and increased the collector's efficiency during the day. In 1909, a new solar heater, called *Day and Night,* was designed with the collector separate from the storage tank, thus allowing the tank to be thoroughly insulated. The collector consisted of a grid of narrow pipes attached to a black metal absorber enclosed within a shallow glass-covered box. As sunlight heated water in the pipes, the warm water rose into the insulated storage tank mounted directly above the collector, a natural thermosiphon. At night, as the collector cooled, the cold water, being denser, remained in the collector and did not rob heat from the storage tank. This system heated water more quickly and provided hotter water longer, and it soon put *Climax* out of business.

These early systems directly heated domestic water—that is, the potable water used within the household. They were simple, easy to maintain, and economical. They were, however, inherently limited to mild climates because even a single night of subfreezing temperature could burst pipes and destroy the collector. To remedy this, in 1913, a new model of the *Day and Night* solar heater added alcohol to the water passing through the pipes in the collector, providing a form of antifreeze. Although the water-alcohol mixture allowed the collectors to experience subfreezing conditions without bursting, it was no longer drinkable and, therefore, could not mix with the domestic water. This created the first indirect system, where the collector fluid and the domestic water remain separate. As the water-alcohol mixture within the storage tank warmed, it conducted heat to the domestic water passing through a coil of copper tubing within the tank, as shown in Figure 6.2. The coiled tubing represents a type of heat exchanger, which is common in many solar hot water systems today.

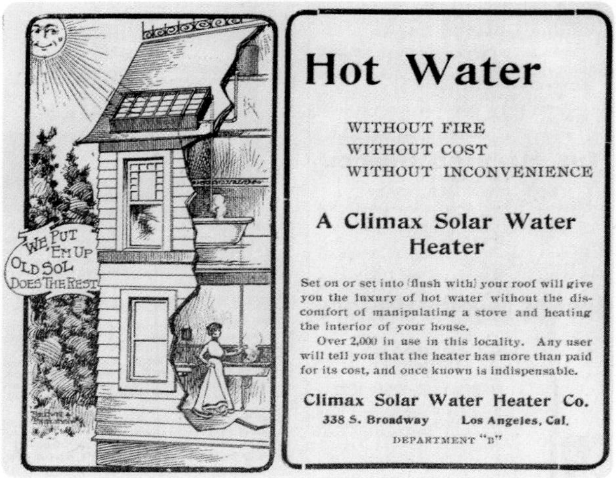

Hot Water

WITHOUT FIRE
WITHOUT COST
WITHOUT INCONVENIENCE

A Climax Solar Water Heater

Set on or set into (flush with) your roof will give you the luxury of hot water without the discomfort of manipulating a stove and heating the interior of your house.

Over 2,000 in use in this locality. Any user will tell you that the heater has more than paid for its cost, and once known is indispensable.

Climax Solar Water Heater Co.
338 S. Broadway   Los Angeles, Cal.
DEPARTMENT "D"

FIGURE 6.1 The world's first commercial solar water heater became available in California in 1891 and sold for $25.

Source: Period Paper.

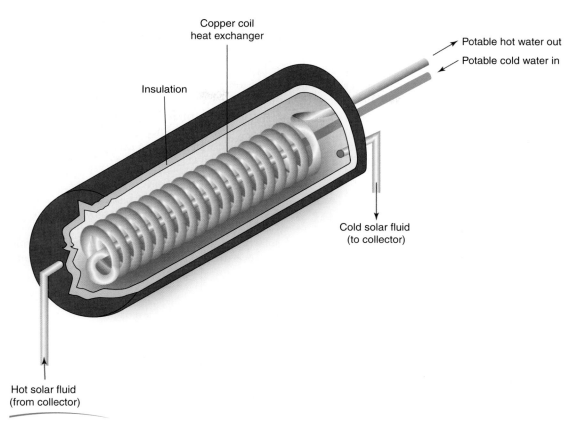

Copper coil
heat exchanger

Potable hot water out

Potable cold water in

Insulation

Cold solar fluid
(to collector)

Hot solar fluid
(from collector)

FIGURE 6.2 The first indirect system transferred heat from sun-warmed fluid in the collector to potable water that flowed through a coil of copper tubing in the storage tank. Modern heat exchangers operate in a similar manner.
© 2016 Cengage Learning®.

The addition of antifreeze to the fluid circulating through the collector (called the heat transfer fluid or solar fluid) expanded the climatic regions suitable for solar hot water systems. This helped popularize hot water systems beyond warm climates and led to improvements in design. By the 1920s, flat plate collectors were available in California and Florida, and the concept soon spread to Japan, Israel, and Australia, where further innovations advanced the technology.

Although the sun offers an inexhaustible supply of free energy, it is rarely able to replace conventional heating entirely because direct sunlight is not always available. Cloudy weather can prevent solar heating for weeks at a time in many regions, making the sun an inherently intermittent resource. Consequently, for most applications, solar hot water systems supplement—rather than replace—conventional systems, a condition that has allowed political and economic influences to severely impact their marketability.

After the initial popularity of solar water heating in the early 1900s, the industry collapsed for years at a time as the availability of inexpensive fossil fuels lowered demand for the sun. Fossil fuels promised hot water any time of the day regardless of weather. Only after the 1973 oil embargo did solar hot water systems establish a solid foothold around the world. Some countries, such as Israel, even passed laws requiring that all new residential construction include solar heating systems. As a result, more than 90% of Israeli households own solar water heaters today.

Modern solar hot water systems include a variety of options applicable to differing climates. Some are relatively unchanged from the earliest designs, while others include pumps, sensors, selective absorbers and coatings, and advanced insulation. This chapter investigates several common types of solar hot water systems used in residential and some commercial applications.

Types of Solar Hot Water Systems

Variations in design have enabled solar hot water systems to accommodate a wide diversity of climates and applications. Nevertheless, two basic classifications serve to distinguish these systems.

1. Passive versus active systems:
   - Passive solar hot water systems rely solely on the thermosiphon effect—where hot water rises and colder water sinks—to transport fluid through the system. They require no pumps or mechanical devices, and their simplicity makes them very reliable, relatively inexpensive, and low maintenance.
   - Active solar hot water systems use electric pumps to move fluid. This added complexity often increases maintenance requirements and operational costs but provides greater design versatility.

2. Open-loop versus closed-loop systems:
   - Open-loop solar hot water systems directly transport domestic water through the collector. Open-loop systems are limited to applications with nonfreezing conditions and perform best using water with relatively low dissolved mineral content, as minerals can precipitate out in the collector and compromise its operation.
   - Closed-loop solar hot water systems circulate a solar fluid through the collectors, rather than domestic water itself, and use heat exchangers to transfer heat from the solar fluid to the domestic water. At no time does solar fluid mix with domestic water. These systems offer greater design versatility than open-loop systems but are also more complex and less efficient.

Variations of these two basic categories have produced numerous types of systems over the last century, many of which have had limited application and have since been replaced with more versatile progeny. The five most common solar hot water systems used in residential applications today include integral collector storage, thermosiphon, open-loop direct, closed-loop pressurized, and drainback systems. These systems are compared in Table 6.1.

Integral Collector Storage

The earliest solar hot water systems directly heated roof-mounted tanks of water and were a type of **integral collector storage (ICS)** system. The darkly painted tanks of an ICS system not only store hot water but also serve as the collector. They are passive, open-loop systems applicable to warm or tropical climates or for seasonal use in vacation homes, campgrounds, or recreational facilities. Since potable water circulates through the collector, the absence of antifreeze limits these systems to above-freezing conditions. When used seasonally, the tanks can be drained and left empty during the winter.

To increase efficiency and extend the usable season, many ICS systems utilize insulated low-emissivity (low-E) glass, selective coatings on the tank itself, internal reflectors, and foam insulation surrounding the tank and pipes, as shown in Figure 6.3. This allows the system to survive brief periods of mild frost. Nevertheless, ICS systems cool considerably at night and during cloudy weather due to the relatively poor insulation of glass. Consequently, they are inherently less efficient than systems that separate the storage tank from the collector.

Because of their simplicity, ICS systems are among the least expensive to install, operate, and maintain. They require nothing beyond a rooftop tank and household plumbing, though many are coupled to backup water heaters, as shown in Figure 6.4.

FIGURE 6.3 Each of these ICS units consists of a selectively coated metal tank enclosed behind low-E glass and secured in an insulated housing. The reflective housing focuses sunlight onto the tank, directly heating potable water. Cold water enters at the bottom and rises to the top, where a second pipe (not shown) delivers heated water for use. The simplicity and low cost of ICS systems make them popular around the world. Courtesy of Premier Solar Technologies LLC.

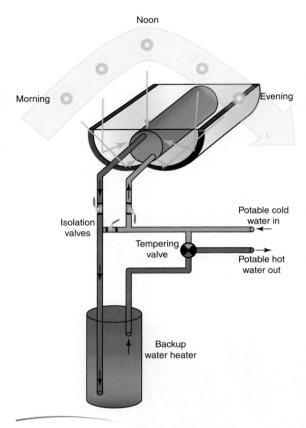

Noon

Morning

Evening

Potable cold water in

Isolation valves

Tempering valve

Potable hot water out

Backup water heater

FIGURE 6.4 ICS system: As potable water warms in the collector, it naturally rises to the top of the tank. This sun-warmed water can either be used directly or be plumbed to a backup water heater, as shown. © 2016 Cengage Learning®.

Whenever hot water is used, water pressure from the building forces cold water into the bottom of the ICS tank, replacing the hot water withdrawn.

In addition to climate limitations, a disadvantage of ICS systems is that they require substantial roof support. Since the collector is also the storage, these systems often weigh over 500 pounds and can cause structural damage to roofs not designed to support such concentrated loading.

Thermosiphon

Thermosiphon systems are the most popular solar water heating system in the world, common in Israel, India, Greece, Australia, and Japan. Similar in operation to the first system developed in 1909, the modern thermosiphon is a simple passive system that consists of a well-insulated tank mounted directly above a collector, as shown in Figure 6.5. As sunlight warms solar fluid in the collector, the fluid rises from the top of the collector into the storage tank, drawing colder fluid into the bottom of the collector. As the collector cools at night, the colder fluid settles to the bottom of the collector, stopping the convective motion of the thermosiphon and preventing heat loss from the storage tank.

Thermosiphons can be either open-loop or closed-loop systems, depending on the climate. In warm and tropical climates, the solar fluid can be potable water, creating an open-loop system. In temperate and cool climates where occasional freezing conditions occur, the solar fluid contains water mixed with antifreeze (typically propylene glycol) and passes through a heat exchanger within the storage tank, creating a closed-loop system. In either case, the movement of heated fluid is passive, requiring no pumps or mechanical apparatus, as shown in Figure 6.6. These systems can be used with any pressurized plumbing system, even in remote areas lacking electricity.

To ensure hot water even on cloudy days, the solar storage tank can be plumbed to a backup water heater that provides additional heat when needed. Like ICS systems, thermosiphon tanks can be quite heavy and require substantial roof support, but their simplicity makes them exceptionally reliable and minimizes maintenance.

Open-Loop Direct

An **open-loop direct system** relies on a small, variable-speed pump to move domestic water through a collector and into a solar storage tank. Open-loop direct systems are the simplest of all active systems but are applicable only to regions that never experience freezing conditions, since domestic water is always present in the collector. Unlike the thermosiphon system, the solar storage tank is mounted below the collector, generally within the conditioned space of the building. This removes the large weight of the storage tank from the roof and reduces heat loss at night. Like the ICS and thermosiphon systems, the solar storage tank of an open-loop direct system may be plumbed to a backup water heater to provide additional heat on cloudy days, as shown in Figure 6.7.

Despite placing the storage tank below the collector, only a very small pump is required to lift water into the collector because an equal amount of water is always descending from the collector. The small pump simply overcomes friction within the system and operates on as little as 10 watts of electricity. This allows a small photovoltaic panel mounted next to the collector to operate the pump. Whenever sunlight illuminates the panel, the pump

FIGURE 6.5 Modern thermosiphon systems consist of an insulated tank placed immediately above a solar collector. Hot water naturally rises into the tank without requiring a pump. © Royster/Shutterstock.com.

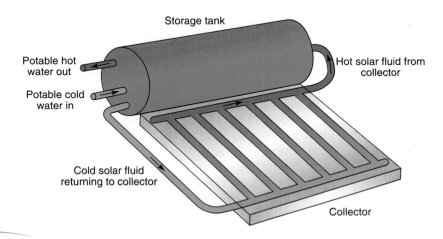

FIGURE 6.6 Thermosiphon system: Cold water enters at the bottom of the collector and rises as it warms in the sunlight. The hottest water ascends to the top of the storage tank, where a pipe makes it available for domestic use. Cooler water near the bottom of the tank sinks back to the collector for additional heating. © 2016 Cengage Learning®.

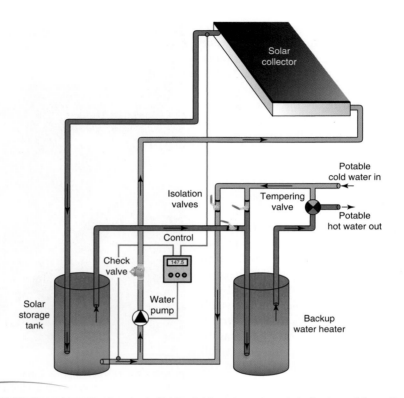

FIGURE 6.7 Open-loop direct system: Cold potable water enters at the bottom of the collector, warms, and exits at the top. An insulated pipe conveys heated water to a solar storage tank. The warmest water rises to the top of the storage tank, where another pipe makes it available for domestic use or plumbs it to a backup water heater, as shown. The small pump required to circulate water can be operated using either a small photovoltaic panel or household electricity that relies on electronic sensors to measure the temperatures of the tank and collector, as shown. © 2016 Cengage Learning®.

turns on and circulates fluid through the collector. On a bright sunny day, the pump runs quickly; on an overcast day, the pump runs slowly; at night, the pump stops. Hence, the use of a photovoltaic panel to operate the pump automatically adjusts the flow rate through the collector according to the solar resource. Although open-loop direct systems are active and require electricity, they can operate in remote regions when utilizing photovoltaic panels.

Alternatively, open-loop direct systems can operate on standard household electricity using a pump and temperature sensors at the collector and storage tank. Whenever the collector is sufficiently warmer than the storage tank, a controller turns on the pump. One advantage of this system, despite the added complexity and reliance on electricity, is that the sensors prevent the pump from circulating fluid through a collector that might be colder than fluid already in the storage tank.

Since the storage tank is located below the collector, hot water within the tank will tend to rise whenever it is warmer than the water above it. This can create a backward circulation during the night—an undesired thermosiphon effect—that allows hot water to circulate through a cold collector. The collector then behaves as a radiator, and this loss of heat can seriously compromise the overall efficiency of the system. To prevent this backward circulation, a one-way valve, called a **check valve**, must be installed on the pipe between the storage tank and the collector. Although relatively inexpensive and low maintenance, check valves represent one of the numerous mechanical devices required of complex systems that may potentially fail and compromise performance.

Open-loop direct systems are the least expensive active system to install. In addition, they require no heat exchanger, use domestic water from standard pressurized plumbing, integrate easily with existing systems, and readily accommodate additional collectors after the original installation.

Although more complex and expensive than ICS and thermosiphon systems, open-loop direct systems offer greater convenience and flexibility of design. The ability to place the solar storage tank at any location within a conditioned space of the building (rather than on the roof) allows for easier maintenance and integration with ordinary household systems. It also allows the storage tank to be better insulated, which increases its efficiency. The primary drawback of this system is its applicability only to nonfreezing locations.

Closed-Loop Pressurized

Closed-loop pressurized systems are a common choice for cold climates. Though relatively complex, they offer fail-safe freeze protection and are extremely versatile, allowing nearly any arrangement of collectors, storage tanks, and other components.

Closed-loop pressurized systems are active, utilizing a small, variable-speed pump to circulate a mixture of water and antifreeze through the collector. A heat exchanger then transfers thermal energy from the solar fluid to the domestic water in the storage tank. Much like the open-loop direct system, the pump circulating the solar fluid can be powered by a small photovoltaic panel adjacent to the collector or operated by household electricity using temperature sensors at the collector and storage tank. The latter is shown in Figure 6.8.

The primary advantage of the closed-loop pressurized system over the open-loop direct system is its ability to tolerate very cold temperatures without freezing. However, the antifreeze solution reduces the efficiency of the system because antifreeze absorbs and stores thermal energy less effectively than pure water. In addition, heat exchangers transfer less energy than would be obtained by direct heating. The reduced efficiency introduced by these measures requires the installation of larger collectors to achieve the same heating potential.

Another limitation of closed-loop pressurized systems is that under prolonged high temperatures the antifreeze can degrade and become acidic, losing its freeze protection and causing serious damage to the collector, pumps, valves, and other components. To avoid overheating the fluid during periods of limited demand for hot water, installers often incorporate a shunt loop that directs hot water through uninsulated pipes in the ground or nearby lake. This cools the solar fluid sufficiently to prevent degradation of the

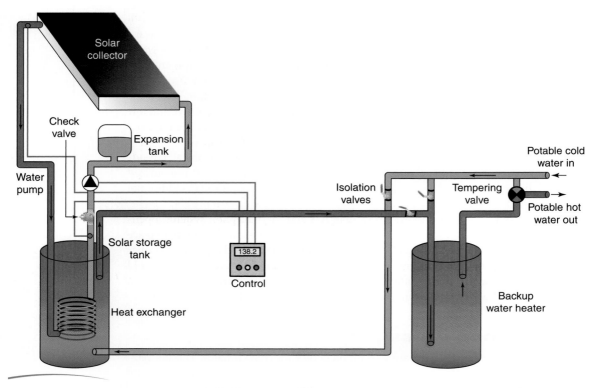

FIGURE 6.8 Closed-loop pressurized system: A heat exchanger transfers energy from the warm solar fluid to the cooler domestic water. The small pump required to circulate solar fluid may be operated by either a photovoltaic panel or household electricity. The hot water from the solar storage tank can be piped directly for domestic use or plumbed to a backup water heater, as shown. © 2016 Cengage Learning®.

antifreeze. Shunt loops are commonly used in residential installations where extended summer absences could otherwise create severe overheating problems.

Even when not overheated, antifreeze gradually degrades and must be replaced every 10 years or so. This requires draining and flushing the collector, heat exchanger, and other components; pumping in new antifreeze; and repressurizing the system.

Closed-loop pressurized systems require an expansion tank as part of the plumbing to allow heated fluid to expand without increasing the system pressure beyond safe levels. As an additional precaution, pressure relief valves are required, as are auxiliary components for filling, venting, and maintaining the system. Furthermore, like the open-loop direct system, the storage tank is located below the collector and requires the installation of a check valve to prevent backward circulation during the night.

Although closed-loop pressurized systems can operate in any climate, they are primarily installed in cold climates, where the protection offered by antifreeze outweighs the added complexity and maintenance requirements and the reduced efficiency.

Drainback

When properly designed, drainback systems operate well in any climate, from the very hot to the very cold, because they can neither overheat nor freeze. Drainback systems are closed-loop active systems that allow solar fluid (typically distilled water) to drain out of the collector and back into a small reservoir, called a drainback tank, when not in use. With no fluid in the collector, the fluid can neither overheat nor freeze.

Drainback systems require temperature sensors at the collector and storage tank. When the collector is sufficiently warmer than the storage tank, a pump circulates solar fluid through the collector, heating it. This fluid returns to the drainback tank and passes through a heat exchanger, where it transfers heat to the storage tank, as shown in Figure 6.9. When the collector is colder than the storage tank or the storage tank has reached its maximum desired temperature, the controller shuts off the pump, and the solar fluid drains back into the reservoir.

Drainback systems are more efficient than closed-loop pressurized systems because they utilize distilled water as the solar fluid rather than an antifreeze solution. With

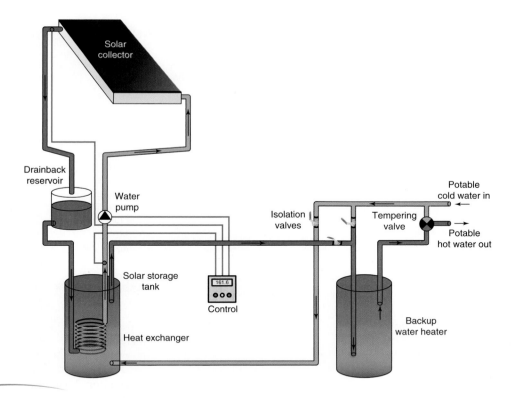

FIGURE 6.9 Drainback system: When the pump shuts off, solar fluid drains from the bottom of the collector into the reservoir, leaving the collector and all exposed pipes empty when not in use. Drainback systems require temperature sensors at the collector and storage tank and utilize larger pumps than the other systems. Hot water can either be used directly for domestic use or be plumbed to a backup water heater, as shown. © 2016 Cengage Learning®.

TABLE 6.1 Comparison of solar hot water systems.

SYSTEM TYPE	CHARACTERISTICS	PASSIVE OR ACTIVE	OPEN-LOOP OR CLOSED-LOOP	ADVANTAGES	LIMITATIONS
ICS	Storage tank also serves as collector. Collector is always filled with water.	Passive	Open-loop	• Simple • Inexpensive • Does not require electricity	• Not suitable for freezing climates • Very heavy; requires substantial support
Thermosiphon	Insulated storage tank is mounted directly above collector. Collector is always filled with fluid.	Passive	Open-loop or closed-loop (depending on climate)	• Simple • Adaptable to different climates • Does not require electricity	• Storage tank must be immediately above collector • Very heavy; requires substantial support
Open-loop direct	Small, variable-speed pump circulates domestic water through collectors and into storage tank. Collector is always filled with water.	Active	Open-loop	• May utilize photovoltaic module for electricity • Integrates easily with existing systems	• Not suitable for freezing climates
Closed-loop pressurized	Small, variable-speed pump circulates solar fluid through collectors. A heat exchanger transfers thermal energy from the solar fluid to the domestic water. Collector is always filled with fluid.	Active	Closed-loop	• Suitable for cold climates • Solar fluid includes antifreeze • May utilize photovoltaic module for electricity	• Antifreeze reduces efficiency • Shunt required to prevent overheating • Greater maintenance
Drainback	Large pump circulates distilled water through collectors. A heat exchanger transfers thermal energy from the distilled water to the domestic water. Collector is empty when not in use.	Active	Closed-loop	• Suitable for cold climates • Utilizes water as solar fluid • Lower maintenance than closed-loop pressurized systems	• Collectors and pipes must drain completely, limiting design options • Large pumps require more electricity to operate • Noisy

© 2016 Cengage Learning®.

no fluid in the collector at night, drainback systems cannot reverse thermosiphon and, therefore, require no check valve. Since these systems are not under pressure, there is no need for expansion tanks, air vents, gauges, or similar mechanical devices that require maintenance.

The relative simplicity of drainback systems makes them reliable, low maintenance, and efficient. They are limited in design, however, because the collector must be located above the drainback tank and must be able to drain completely. This requires tilting the collector and providing sufficient continuous slope in the piping that water cannot stagnate at any point.

The most significant drawback of drainback systems is that they require relatively large pumps to lift water to the collector. Unlike open-loop direct and closed-loop pressurized systems, where all pipes remain constantly filled with water, drainback systems have air in the pipes, eliminating some of the gravitational benefit of the descending water that returns from the collector. These pumps require 150–250 watts of power, depending on the height of the collector above the drainback tank, and this power is usually supplied continuously during operation. Photovoltaic panels are generally inadequate because the pumps require maximum power from the onset and cannot tolerate fluctuations resulting from intermittent cloud cover. Once the pipes are completely filled with water, more advanced drainback systems automatically reduce the power supplied to the pump, conserving electricity.

Drainback systems can also be noisy, sounding like a gurgling fountain or coffee percolator, due to the presence of air in the drainback tank. Nevertheless, the simplicity and low maintenance of drainback systems make them popular across many different climates.

Types of Collectors

Except for the ICS system, which incorporates its own collector, each of the other four hot water systems utilizes some form of collector. The two most common types of collectors are the flat plate collector and the evacuated tube collector. Collectors and mounting options are compared in Table 6.2.

Flat Plate Collector

Flat plate collectors are available in hundreds of different models, most of which consist of a shallow, insulated metal box covered with tempered glass that encloses tubing attached to absorber plates. The absorber plates, usually made of thin sheets of copper or aluminum, are darkly colored or painted with selective coatings that minimize heat loss through radiation. The plates absorb the sun's energy and transfer heat to the solar fluid passing through the collector. Factors that influence the collector's efficiency include the manner in which the absorber plates are bonded to the tubing, the configuration of the tubing, the coatings on the absorber plates, the type of glass, the amount of insulation, and the materials and quality of manufacture.

Flat plate collectors make use of three common tubing configurations:

- **Traditional:** Fluid flows through a series of vertical tubes between a bottom header (where cold fluid enters) and a top header (where warm fluid exits). This maximizes the amount of cold fluid exposed to sunlight and absorbs the greatest amount of solar energy.
- **Serpentine:** Fluid flows through a single continuous tube that meanders from the bottom of the collector to the top, much like an undulating garden hose placed in the sunshine. This warms a single stream of fluid for a longer time, maximizing the output temperature but not the total energy absorbed.
- **Flooded:** Fluid completely floods the collector, entering at the bottom and flowing upward, operating much like the traditional configuration but providing greater absorption area.

The most common configuration for residential hot water systems is the traditional, shown in Figure 6.10. Serpentine configurations are more likely to be used in regions with ample solar resource where very hot water is required or for commercial applications employing many collectors connected in parallel. Flooded configurations may be slightly more efficient than traditional configurations and may become more common as polymers replace metal and glass construction.

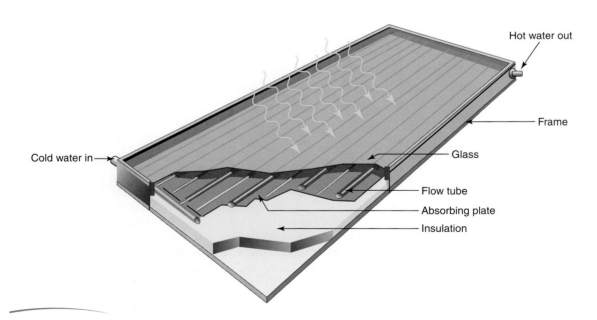

FIGURE 6.10 A typical flat plate collector consists of fluid-filled tubes bonded to dark absorber plates and secured within an insulated glass-covered box. This flat plate collector uses the traditional pipe configuration. © 2016 Cengage Learning®.

The recent introduction of collectors made of flexible polymers promises to provide an economical, freeze-safe, and lightweight alternative to traditional flat plate collectors. Polymer collectors may eliminate the need for antifreeze altogether and create greater design versatility.

Evacuated Tube Collector

Evacuated tube collectors consist of numerous sealed glass tubes attached to a common top pipe called a manifold. Each glass tube encloses an absorber fused to a central heat pipe. Air has been removed from the glass tubes, creating a vacuum that surrounds the absorbers and prevents heat loss by conduction and convection, behaving much like a thermos bottle. In most evacuated tube collectors, solar fluid does not directly circulate through the glass tubes. Rather, the central heat pipe within each tube contains a small quantity of alcohol or similar fluid and is sealed under a partial vacuum. The reduced pressure causes the fluid to vaporize at a lower temperature than normal. When sunlight illuminates the absorbers, the fluid boils, and the hot vapor rises to the top of the heat pipes, where heat exchangers mounted in the manifold transfer energy from the hot vapor to the solar fluid flowing by. As the vapor gives up its heat and cools, it condenses and flows back down the heat pipe, where the process continues. (See Figure 6.11 and Figure 6.12.)

The absence of solar fluid within the glass tubes themselves allows each tube to be replaced if damaged without draining the entire system. Typically, each evacuated glass tube plugs directly into the top manifold and can be easily installed and removed. One limitation of evacuated tube collectors is that the condensing vapor must be able to flow back down the heat pipe, requiring that the collector be mounted in a tilted position (Figure 6.13).

Performance Comparison

Flat plate and evacuated tube collectors perform slightly differently and present different advantages and limitations.

Flat plate collectors are simpler to construct, have lower embodied energy, and typically cost half as much as evacuated tube collectors. They are durable, require very little maintenance, and can be expected to last 50 years or longer. However, flat plate collectors perform optimally only in bright sunshine and only when they receive direct illumination. Sunlight striking the collector surface obliquely tends to reflect from the glass covering. Despite selective coatings on the glass and absorber plates and insulation on the back and perimeter of the collector, flat plate collectors lose considerably more heat to the environment than the vacuum-insulated evacuated tube collectors. On partially overcast days, the heat lost by flat plate collectors may approach that of the heat gained, resulting in minimal, if any, heating. On the other hand, this lesser insulation helps shed snow that may otherwise prevent operation altogether.

The greater insulation level of evacuated tube collectors allows them to more easily attain higher temperatures, particularly in very cold conditions. The curved nature of the tubes also permits them to capture more oblique sunlight, extending the length of time during the day when they gather energy. Evacuated tube collectors also perform better on overcast days because they lose less heat. However, evacuated tube collectors

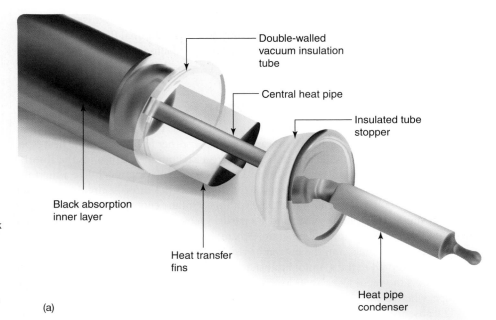

Double-walled
vacuum insulation
tube

Central heat pipe

Insulated tube
stopper

Black absorption
inner layer

Heat transfer
fins

Heat pipe
condenser

(a)

FIGURE 6.11 (a)
This evacuated tube
collector uses a black
cylindrical absorber
and metal fins to
transfer energy to a
central heat pipe.

© 2012 Cengage Learning.
From Steeby/Alternative Energy:
Sources and Systems.

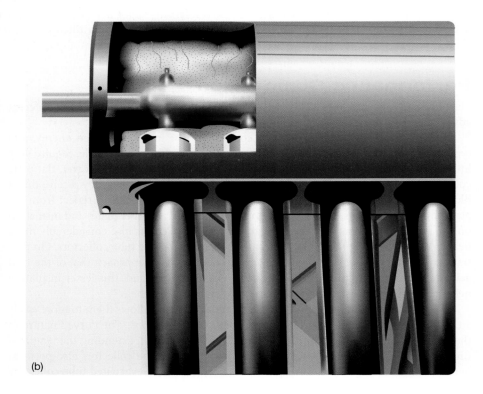

(b)

FIGURE 6.11 (b)
An insulated manifold
transfers heat from
the bulb-shaped
condenser at the end
of the evacuated tube
to domestic water
flowing through.

© 2012 Cengage Learning.
From Steeby/Alternative Energy:
Sources and Systems.

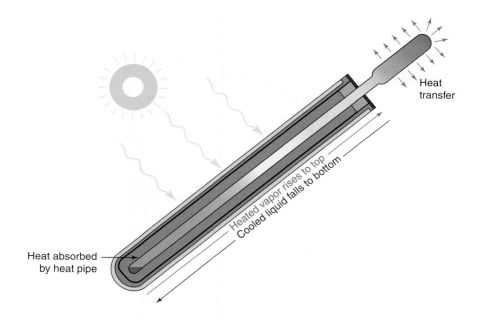

FIGURE 6.12 Each evacuated tube contains a sealed heat pipe filled with a volatile fluid that boils when heated by the sun. As the hot vapor rises and condenses, it transfers heat to the domestic water flowing past. In order for the condensed fluid to flow back down the heat pipe, the collector must be mounted in a tilted position. © 2016 Cengage Learning®.

FIGURE 6.13 These evacuated tube collectors, located in northern Wisconsin, are tilted to maximize winter absorption and shed snow. Courtesy of Scott Grinnell.

use annealed glass, which is much more fragile than the tempered glass of flat plate collectors, making evacuated tube collectors more prone to failure. In addition, due to the spaces between tubes, evacuated tube collectors offer less sunlight-absorbing surface area than similarly sized flat plate collectors. As a result, when comparing equally sized installations, flat plate collectors may outperform evacuated tube collectors. In snowy regions, evacuated tube collectors may allow snow to accumulate around and on top of the collectors, reducing winter performance. This is particularly true in locations receiving heavy, wet snow that may refreeze around the tubes.

Mounting Options

Solar hot water systems utilize rigid pipes to transport solar fluid and must be mounted in a fixed position. They cannot be adjusted to maximize seasonal (or daily) absorption. Securely mounting solar collectors is essential because any sagging or shifting can break pipes or create areas where fluid can stagnate, which in some climates may result in freezing. Three mounting systems provide a variety of options: roof mount, ground mount, and awning mount.

- **Roof mount system**: Collectors mounted on the roofs of buildings are generally held in place by brackets that either tilt the collectors at an optimal angle or hold the collectors parallel to and a few inches above the roof. Elevating the collectors off the roof itself is essential to allow proper drainage and ventilation of the roof. (See Figure 6.14.)
- **Ground mount system**: Ground mount systems can be as simple as four posts anchored in the ground that support and tilt the collector to optimize its performance. (See Figure 6.15.)

FIGURE 6.14 These two flat plate collectors heat water for a drainback system. Cold water enters each panel at the bottom, and hot water exits from the top. Courtesy of Scott Grinnell.

FIGURE 6.15
Ground-mounted systems allow for simpler maintenance. This system provides hot water for a home in Montana. Courtesy of Joe Rightmyer.

- **Awning mount system:** An awning mount system attaches the collector to a vertical exterior wall using horizontal brackets. Longer bottom brackets push the base of the collector away from the wall to achieve the desired tilt. Awning mount systems are often used at high-latitude locations where optimal winter heating requires tilts greater than 60°. Awning mount systems are also commonly placed on the sides of buildings that do not have a south-facing roof, as shown in Figure 6.16.

When evaluating the advantages and limitations of the different mounting options, the following factors should be considered:

- The degree of shading at each site resulting from trees, utility poles, buildings, and so forth
- The proximity of the collector to the point of use; the shorter the pipe run, the less heat lost
- Potential ground-based hazards, including possible vandalism
- The ease of maintenance, including snow removal if applicable
- Material components of the building and future renovations, including replacement of roofing or siding
- Design constraints, such as the weight of an ICS systems, the draining capability of a drainback system, or the likelihood of the collector interfering with or obstructing passive solar features of the building
- Overall expense and ease of installation

Table 6.2 summarizes the advantages and limitations of flat plate and evacuated tube collectors and evaluates the benefits and shortcomings of the three mounting options.

FIGURE 6.16 This awning mount system uses two flat plate collectors to heat water for a home in Hudson, Massachusetts.

Courtesy of Mark Durrenberger/ New England Clean Energy, LLC.

Sizing Solar Hot Water Systems

The available solar resource varies, depending on weather conditions. Normal changes in cloud cover make solar heating intermittent, so sizing a system to provide all of a building's domestic hot water needs is usually impracticable. In regions that experience prolonged periods of cloudy weather, solar hot water systems require some form of backup heat. Conversely, in regions that experience only occasional cloudy weather, solar hot water systems can provide all of the domestic requirements simply by installing a sufficiently large, well-insulated storage tank. In general, most solar hot water systems rely on conventional backup sources to ensure an uninterrupted supply of hot water.

Variations in solar resource by season, unlike those resulting from weather, are predictable and can be accommodated. For year-round use, collectors can be tilted to maximize winter absorption (typically at an angle equal to the latitude plus 15°, as discussed in Chapter 5), which reduces summer heat gain and can help prevent overheating, particularly for closed-loop pressurized systems.

TABLE 6.2   Collectors and mounting options.

	ADVANTAGES	LIMITATIONS
Collector		
Flat plate	• Lower embodied energy • Less expensive • Minimal maintenance • Better at shedding snow	• Requires direct illumination • More heat loss to environment • Lower water temperature
Evacuated tube	• Better insulated • Attains higher temperatures • Captures more oblique sunlight	• More fragile construction • May not shed snow well
Mounting		
Roof	• Minimizes shading • Minimizes heat loss • Allows the use of drainback systems • Removes the collectors from ground-based hazards • Frees up yard space	• Maintenance is more difficult • Must be removed to reroof building • ICS and thermosiphon systems may be too heavy
Ground	• Permits easy maintenance • Accommodates heavy systems such as ICS and thermosiphon • Does not impair ability to replace roofing	• More expensive • May experience more shading • More vulnerable to ground-based hazards • Does not allow the use of drainback systems • Greater heat loss between the collector and building
Awning	• Minimizes shading • Minimizes heat loss • May permit the installation of a drainback system • Easily sheds winter snow • Frees up yard space	• May interfere with windows and passive solar features • ICS and thermosiphon systems may be too heavy

© 2016 Cengage Learning®.

The appropriate size of a solar hot water system can be calculated by considering the collector performance, the available solar resource (given in Figure 5.19), and the household's hot water demands. The Solar Rating and Certification Corporation (SRCC), established in 1980, measures and certifies the performance of solar collectors and provides useful information on expected heat output. In general, the more sunlight a region receives, the more water can be heated by a square foot of collector.

Figure 6.17 roughly groups regions within the United States by their solar resource and provides an estimate of the number of gallons of water per square foot of collector that can be heated from the mean ground temperature to the standard 120°F used in most domestic systems. For example, in the desert climate of the Southwest, each square

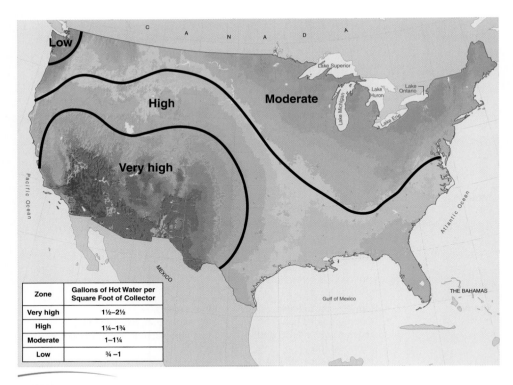

Zone	Gallons of Hot Water per Square Foot of Collector
Very high	1½–2½
High	1¼–1¾
Moderate	1–1¼
Low	¾–1

FIGURE 6.17 The Southwest provides the greatest solar resource within the United States. Each square foot of collector surface is roughly capable of heating between 1½ and 2½ gallons of water from the local ground temperature to the desired 120°F. The Pacific Northwest, by comparison, requires more than twice the collector surface to meet the same heating needs. © 2016 Cengage Learning®.

Source: Adapted from the National Renewable Energy Laboratory (NREL).

foot of collector can heat as much as 2½ gallons of hot water per day. In the Pacific Northwest, which is often overcast, each square foot of collector can heat only ¾ of a gallon of hot water per day on average. These values offer a method to estimate appropriate collector size.

The average American household consumes about 20 gallons of hot water per person per day. This number can vary considerably, depending on the efficiency of household appliances and the habits of the occupants. Front-loading, high-efficiency clothes washing machines typically use only a third as much hot water as conventional top-loading machines. A household with small children requiring daily baths will generally use more hot water than one with individuals taking short showers. Nevertheless, 20 gallons of hot water per person per day provides a helpful benchmark for designing hot water systems. For example, an average family of four will require an 80-gallon hot water storage tank with a total collector area of between 32 square feet (in Yuma, Arizona) and over 100 square feet (in Seattle, Washington).

Wisconsin Home

Location: Northern Wisconsin (46.5° N latitude)

Type of System	Collector Type	Collector Size	Primary Usage	Date of Installation
Drainback	Flat plate (Heliodyne)	Two panels, each 27 square feet (54 square feet total)	Domestic hot water	2011

Description of System: For year-round use, the cold climate of northern Wisconsin limits the selection of solar hot water systems to two choices: (1) the closed-loop pressurized system and (2) the drainback system.

Each of these two systems has advantages and disadvantages. Originally, the owners considered a ground-mounted collector, a choice that would have required the use of a closed-loop pressurized system. However, a site assessment using a solar pathfinder indicated that a ground-mounted system would experience shading in the winter. Similarly, the passive solar design of the house (with numerous south-facing windows protected by large eaves) prevented the use of an awning mount, as the collectors would shade the

Courtesy of Scott Grinnell.

windows and potentially be damaged by snow sliding from the eaves. Consequently, the owners decided to mount the collectors on the roof and utilize a drainback system.

A drainback system eliminated the need for a shunt loop to prevent overheating in the summer, reduced maintenance, and increased the efficiency of the collector by using distilled water for the solar fluid rather than an antifreeze solution. However, the drainback system increased electricity consumption by requiring a much larger pump.

The owners sized the collectors so that a sunny day would provide enough energy to heat 80 gallons of water from an initial temperature of 40°F to a maximum tank temperature of 160°F. This quantity of water is roughly triple their daily domestic water needs and prevents the use of the electric backup water heater except during conditions of prolonged cloudiness. The collectors were oriented due south and tilted at 60° to maximize winter efficiency. The steep angle also facilitated snow removal. The abundant sunlight during the summer provides ample hot water despite the steep tilt.

Cost Analysis: The system satisfies nearly all of the domestic hot water needs for the household from May through October and roughly 30% of the hot water needs during the remainder of the year, when winter cloud cover sometimes persists for weeks at a time. The system offsets the cost of operating a conventional electric water heater and saves approximately 3700 kWh of electricity per year. At $.12/kWh, the savings amount to $440/year, a sum that will likely increase as energy becomes more expensive.

The installation of the system cost $5450, including a federal tax incentive that paid for 30%. This represents more than an 8% return on investment, and the system will pay for itself in slightly over 12 years.

Even in the cold and often cloudy climate of northern Wisconsin, solar hot water systems offer a cost-effective means of harnessing natural renewable energy. Day after day, these systems provide reliable hot water and a sense of stability amid ever-increasing energy prices.

Industrial Hot Water Systems

Although the first commercially available solar water heaters were designed for residential use, industrial applications had their origins around the same time, with the first solar-powered printing press operating by 1882, as shown in Figure 6.18. Industrial systems are different from residential systems because they typically demand hotter water than can be obtained by flat plate or evacuated tube collectors. Instead, industrial systems often use parabolic trough collectors to concentrate sunlight on fluid-filled pipes. These pipes can become significantly hotter than the boiling temperature of water and are typically filled with oil.

Modern applications of solar heating for industrial purposes include water treatment and desalination, food and beverage processing, pulp and paper production, textile washing and dyeing, chemical distillation, pharmaceuticals, and mining and quarrying.

Although different in terms of technology, industrial-scale hot water systems operate on the same basic principles as residential systems. (See Figure 6.19.)

FIGURE 6.18 In 1882, a steam-powered printing press became one of the first industrial applications of solar hot water. The parabolic reflector focused the sun's rays onto a water-filled chamber to generate steam. © 2013 Cengage Learning. From Hinrichs&Kleinbach/Energy, 5E.

FIGURE 6.19 This array of tracking parabolic troughs provides hot water for the Jefferson County Jail near Golden, Colorado. The hot water is used for the kitchen, bathing, and laundry facilities. *Source:* Warren Gretz/NREL.

Chapter Summary

Solar hot water systems represent one of the most cost-effective means of harvesting natural renewable energy. They are usually the first to be implemented in residential green building projects and in many climates pay for themselves in less than 5 years.

Simple, reliable, and with very little maintenance, solar hot water systems can provide freedom from the expense and environmental degradation of fossil fuel consumption, bestow comfort and an improved quality of life on those who lack conventional resources, and impart a deep satisfaction to homeowners by harnessing a natural and inexhaustible resource.

Although different climates call for different types of hot water systems, all inhabited regions can make use of this simple technology and benefit from the free energy offered by the sun.

Review Questions

1. What type of solar hot water system would you recommend to heat domestic water for a three-bedroom home in Hawaii? Explain your reasoning.

2. The solar fluid used in closed-loop pressurized systems (usually a mixture of 50% propylene glycol and 50% water) absorbs and transports 15% less heat than pure water. In light of this inefficiency, explain why closed-loop pressurized systems are often selected over drainback systems, which utilize pure water.

3. A yurt located in the remote mountains of Washington is used heavily during the summer by up to eight backpackers per night. The caretaker wishes to install a system capable of providing adequate hot water for all occupants during the summer. Although no electricity is available, a nearby stream fills an elevated tank to provide pressurized water. The yurt is unheated and, therefore, not used during the winter.
   a. What type of solar hot water system would you recommend for the yurt? Explain your reasoning.
   b. What type of mount (roof, ground, or awning) would you suggest for the collectors? Why?

4. Prior to mounting solar collectors on the roof of an old residential building, the owner decides to remove the existing asphalt shingles and reroof the building.
   a. Why is this a good idea?
   b. What type of roofing would you recommend?

5. Design a solar hot water system for year-round use in a four-bedroom house located in a cold climate that receives substantial snowfall. Include the type of system, choice of collector, and method of mounting. Justify your selections.

6. Evacuated tube collectors must maintain a minimum tilt angle of 20° so that condensing fluid inside the heat pipe can adequately drain back to the absorber. In what regions of the world could this requirement limit achieving the optimal tilt angle for maximum efficiency?

7. Explain how each of the following considerations affects the time required for a solar hot water system to pay for itself by offsetting conventional heating costs:
   a. Locally available incentives that offer funding for renewable energy installations
   b. Cost of conventional fuel
   c. Hot water consumption habits of the building's inhabitants
   d. Installation of low-flow fixtures and other conservative measures

Practice Problems

1. At a residence in Sedona, Arizona, each square foot of a flat plate collector absorbs enough solar energy to heat 2 gallons of domestic water from an initial temperature of 61°F (the mean ground temperature) to the desired temperature of 120°F. Determine the number of 4- × 8-foot collectors required for a household of three people, assuming an average water consumption of 20 gallons per person per day.

2. A household located in International Falls, Minnesota, where the mean ground temperature is only 38°F, is occupied by four people. At this location, each square foot of flat plate collector gathers enough solar energy to heat only 1 gallon of domestic water on a sunny day.
   a. Determine the number of 4- × 8-foot collectors required to meet the hot water needs of the household, assuming average water consumption. Consider only whole numbers.
   b. What maximum hot water consumption per person would allow the use of one less solar collector?

3. One British thermal unit (Btu) is defined as the energy required to raise the temperature of 1 pound of water by 1 degree Fahrenheit. Since each gallon of water weighs 8 pounds, it takes 8 Btu to warm 1 gallon of water by 1°F.
   a. How many Btu are required to raise 80 gallons of 38°F water to 120°F?
   b. If an evacuated tube collector gathers 850 Btu/ft^2 during a cold clear day in northern Minnesota, what total collector area is required to heat the 80 gallons?

4. Pipes carrying water between a ground-mounted solar hot water collector and a nearby building must be insulated to minimize heat loss. This insulation typically has an R-value of only 2.5. Collectors positioned 20 feet from a building require about 50 feet of piping, and the total surface area of the pipe is about 11.5 square feet.
   a. If 140°F water flows through a pipe buried in ground at 50°F, determine the rate of heat loss (in Btu/hr).
   b. If the collector absorbs 5400 Btu/hr of solar energy, what percentage is lost in the buried pipes?

Solar Electricity

CHAPTER 7

Source: Warren Gretz/NREL.

Introduction

The sun's energy can be utilized not only to warm buildings through solar design (Chapter 3) and create hot water for domestic and industrial use (Chapter 6) but also to create electricity.

Electricity is the foundation for much of technology and essential for the structure of modern society. Interestingly, all commercial electricity is produced by the same basic technique—by rotating the shaft of a generator. Conventional and nuclear power plants create high-pressure steam to spin a turbine connected to a generator. Wind turbines use the kinetic energy of moving air to rotate the shaft of a generator. Hydroelectric plants use the kinetic and gravitational energy of falling water to rotate the shaft of a generator. Ocean current and tidal power plants use the motions of the sea to rotate the shaft of a generator. Although the engineering varies somewhat among energy sources, the basic principles for producing electricity remain the same. However, one form of electricity production—the direct conversion of sunlight into electricity, called **photovoltaics (PV)**—is fundamentally different. Unlike any other form of power generation, photovoltaic electricity involves no moving parts, requires no lubrication or routine maintenance, and is utterly silent.

The photovoltaic effect was first noted by Edmond Bacquerel in 1839 and later explained by Albert Einstein in 1905. In 1918, a method for manufacturing single crystals of silicon was discovered, but it was not until the 1950s, with work at Bell Laboratories, that its value for the production of solar cells became apparent. Spurred by intensive research for the space program, the first silicon solar cell was created in 1958 and used for the Vanguard I satellite. The technology became commercially available soon afterward, but it was not until the 1970s that large companies began producing usable photovoltaic modules. Sharp Corporation created the first solar calculator in 1978. Today photovoltaic technology infuses watches, calculators, and cell phone chargers; provides power in remote areas for pumping water, operating navigational and telecommunication systems, and maintaining refrigeration, sanitation, and health care; and has found widespread use in residential and industrial applications (see Figure 7.1). As one of the more glamorous forms of technology in the energy field, photovoltaic modules have become iconic of renewable energy systems.

FIGURE 7.1a Photovoltaic modules provide power for many remote applications where the utility grid is unavailable. This array, used as part of a runway lighting system in Antarctica, guides cargo planes to a safe landing on the ice. *Source:* Northern Power Systems/NREL.

FIGURE 7.1b Portable electronics benefit from photovoltaic technology. An hour or two of sunlight can recharge the batteries of this computer. *Source:* John Lenz/NREL.

FIGURE 7.1c Photovoltaic modules integrated into carports create electricity silently and unobtrusively and may soon be common for recharging electric vehicles. *Source:* Sandia National Laboratories/NREL.

FIGURE 7.1d In space, where other forms of energy are severely limited, sunlight provides ample, reliable power. The International Space Station employs over 32,000 square feet of photovoltaic modules and generates more than 85 kW of electricity, the largest energy collection system ever put into space. *Source:* NASA.

Although photovoltaic technology represents the most direct conversion of solar energy into electricity, it is not the only form of solar electricity. Solar thermal power plants also produce electricity by focusing the sun's energy to boil water and create steam, which then drives turbines in much the same manner as conventional power plants. Although this technology can be utilized at the residential scale, it is primarily employed for commercial and industrial applications. This chapter will investigate these two forms of solar electricity.

The Photovoltaic Effect

The process of converting sunlight directly into electricity occurs at the microscopic scale. The inability to witness this phenomenon visually has heightened its mystique ever since its first detection. Electricity is the coherent motion of electrons, and the discovery that sunlight could move electrons was both happenstance and profound.

When a material absorbs the electromagnetic energy of light, the energy excites electrons from their usual positions in atoms to higher energy states, where they are more able to move around. Ordinarily, the excited electrons quickly relax back to their ground states, giving up the absorbed energy as heat or re-emitted light. A **photovoltaic device** is one in which some built-in asymmetry draws the excited electrons away and delivers them to an external circuit before they can relax.

This asymmetry can be created in a type of material called a **semiconductor**. A semiconductor is a nonmetallic material, such as silicon, that ordinarily allows electrons very limited mobility. Unlike metals (which permit electrons to move freely) and insulators (which almost completely prevent the flow of electrons), semiconductors allow electrons to move freely only when the electrons are in the excited state. Joining two thin layers of dissimilar semiconductors creates the type of asymmetry needed to produce the photovoltaic effect.

Early photovoltaic materials operated at very low efficiency, converting 1% or less of the available solar energy into electricity. Improvements depended on optimizing the two dissimilar semiconductors. The most commonly used semiconductor is silicon, a primary component of sand and one of the most abundant naturally occurring elements in the earth's crust. Silicon is used for both layers, which are made dissimilar by deliberately diffusing minute quantities of carefully selected impurities, notably phosphorous and boron, into extremely pure silicon crystals. This process, known as **doping**, dramatically alters the electrical behavior of silicon. The layer doped with phosphorous introduces extra electrons into the silicon crystal, since each phosphorous atom contains one more electron than the silicon atom it replaced. These extra electrons are not bound by the crystal structure and are able to move freely. This layer is referred to as an **n-type (negative) semiconductor** due to the presence of extra electrons. The layer doped with boron, on the other hand, creates a deficit of electrons, since each boron atom has one less electron than the silicon atom it replaced. This layer is referred to as a **p-type (positive) semiconductor** due to the deficit of electrons.

Initially both the n-type and the p-type layers are neutral (despite their nomenclature) because the atoms of phosphorous and boron are both neutral. Any surplus or deficiency of electrons introduced into the silicon crystal is balanced by an equal surplus or deficiency of protons in the nuclei of the doped atoms. However, joining these two dissimilar layers together causes some of the surplus electrons in the n-type semiconductor to migrate across the boundary (known as a **p-n junction**) and fill the deficiencies in the p-type semiconductor. This relocation of electrons sets up an electrical imbalance across the junction, with the p-type semiconductor acquiring a net negative charge and the n-type becoming positive. (See Figure 7.2.)

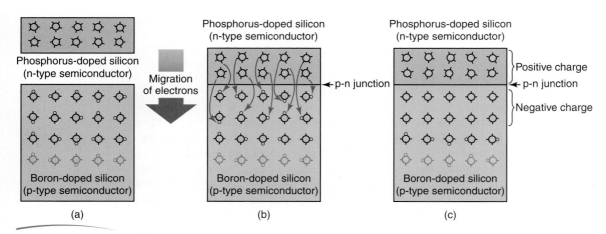

FIGURE 7.2 (a) When silicon is doped with phosphorous to produce an n-type semiconductor, each atom of phosphorous introduces an extra electron (shown as -) that is free to move about the crystal. When silicon is doped with boron to produce a p-type semiconductor, each atom of boron introduces a shortage of one electron (shown by o) that acts as a hole into which free electrons may fall. Both semiconductors are electrically neutral, since each has the same number of protons as electrons. (b) When an n-type semiconductor is joined to a p-type semiconductor, the extra electrons in the n-type semiconductor migrate across the p-n junction to fill the holes in the p-type semiconductor. (c) This creates an imbalance of charge in both layers: the addition of electrons creates a negative charge in the p-type semiconductor, and the removal of electrons creates a positive charge in the n-type semiconductor. This electrical imbalance tends to drive electrons excited by sunlight up to the top of the n-type semiconductor. © 2016 Cengage Learning®.

As sunlight strikes a solar cell, electrons near the p-n junction absorb some of the energy and jump into an excited state, where they are able to move freely. The electrical pressure created by the imbalance of electrons drives the free electrons from the p-n junction toward the surface of the n-type semiconductor. Thin metal conductors draw the electrons into an external circuit and deliver them back to the p-side, where they return to their ground state. The different electrical characteristics imposed by the doped phosphorous and boron atoms continuously drive electrons across the p-n junction and maintain the imbalance needed to continue this process indefinitely.

Silicon Solar Cells

A **silicon photovoltaic cell** consists of a thin layer of n-type semiconductor joined to a thicker substrate of p-type semiconductor, shown in Figure 7.3. The top surface of the n-type semiconductor is textured and coated to reduce reflection. Thin metal strips attached to or etched into the surface of the n-type semiconductor gather electrons that migrate up from the p-n junction and direct them through an external circuit. A rear metal backing on the p-side completes the circuit, delivering electrons to the awaiting vacancies within the p-type crystal.

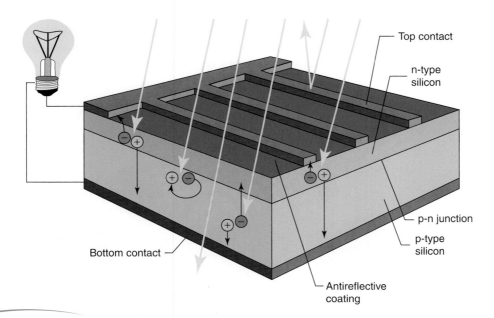

FIGURE 7.3 When sunlight strikes a solar cell and gives up its energy to an electron, the liberated electron leaves behind a hole. Much like moving a parked car to fill a more desirable space closer to an attraction, a nearby electron will move into the hole if the hole is closer to the top of the cell. The electron that moved also leaves behind a hole, which will be filled by an electron even farther from the top, and so on. In this way, the hole slowly sinks to the bottom of the cell. Electrons collected by the top contact flow through an external circuit (perhaps lighting a bulb) and return to the bottom contact, where they fill the awaiting holes. As long as sunlight continues to energize the cell, electrons continuously migrate to the top, and holes continuously sink to the bottom, establishing a current. Occasionally, a liberated electron will fall back into its own hole, a process called *recombination,* which reduces the cell's efficiency. © 2016 Cengage Learning®.

Different manufacturing methods create three different types of silicon cells—monocrystalline, polycrystalline, and amorphous—each with slightly different properties.

Monocrystalline Cells

Monocrystalline solar cells begin with a single crystal of silicon, one having virtually no defects or impurities. The crystal grows around a rod that is slowly pulled from a tub of molten silicon, producing a cylindrical column that can range from 7 to 14 inches in diameter and several feet in length, as shown in Figure 7.4.

This crystal is sliced into extremely thin wafers and doped with phosphorous and boron, creating the n-type and p-type semiconductors. These are joined together to create the p-n junction, and the wafers are polished and sometimes trimmed to better cover a rectangular module. Nevertheless, to reduce waste, most modules made from monocrystalline silicon have gaps around the corners of each cell. Since these gaps do not participate in the production of electricity, they reduce the active area of the module. Monocrystalline silicon cells are black or dark blue in appearance due to the antireflective coating, as shown in Figure 7.5.

Monocrystalline cells are the most efficient of all silicon cells, converting 15%–20% or more of the arriving solar energy into electricity. They are durable and very reliable, producing power year after year, though gradual degradation reduces their efficiency. However, this degradation is small—generally less than 0.5% per year (depending on the circumstances)—due to the very stable crystal structure. Monocrystalline cells created 40–50 years ago are still in use, and many are producing power with less than a 10% decline in efficiency. Monocrystalline cells are the most expensive to manufacture

FIGURE 7.4 A single crystal of silicon grows into a long cylinder as it is rotated and drawn from a bath of molten silicon. iStockphoto.com/Coddy.

and contain the greatest embodied energy, since achieving a single pure crystal of silicon is energy and labor intensive.

Polycrystalline Cells

Polycrystalline solar cells are made from molten silicon that is poured into a mold, cooled, and sliced into thin wafers. The wafers are then polished and doped to create the p-n junction. As the silicon cools in the mold, it develops into many different crystals, creating a multifaceted appearance like polished granite (see Figure 7.6). Although polycrystalline cells are simpler to manufacture than monocrystalline cells, the boundaries between crystals inhibit electron migration through the semiconductor and reduce the overall efficiency of the cell to 12%–15%. Polycrystalline cells do not need to be trimmed to fully cover a photovoltaic module and are, therefore, more resource efficient than monocrystalline cells. The simplified manufacturing process also reduces their embodied energy and expense. Polycrystalline cells degrade slowly over time but can be nearly as stable as monocrystalline cells.

FIGURE 7.5 Individual monocrystalline solar cells are often trimmed to better cover a rectangular module. The waste silicon can be recycled, but the process is energy intensive. Many fine metal lines gather electrons and pass them on to the two thicker wires, which send them through an external circuit. Silicon solar cells appear blue or black due to an antireflective coating. *Source:* Rick Mitchell/NREL.

FIGURE 7.6 The many small crystals comprising polycrystalline silicon reflect light slightly differently and give the finished cell a multifaceted appearance. Polycrystalline cells can be cast in square molds and require considerably less trimming, reducing labor and energy. However, the numerous crystal boundaries reduce the cell's efficiency. Courtesy of Scott Grinnell.

Very thin polycrystalline ribbons can be drawn directly from molten silicon. This method eliminates the need to saw the silicon into thin wafers and saves material and energy, reducing costs. However, the crystals tend to be smaller, introducing even more boundaries and reducing efficiency even further.

Amorphous Cells

Amorphous solar cells (Figure 7.7a) use noncrystallized silicon that is deposited in very thin films directly onto a substrate material. They are typically only 5%–8% efficient but offer greater versatility. They can be manufactured in nearly any shape and size with minimal waste, made to be semitransparent and laminated on glass to harness energy while reducing window glare, adhered to metal roofing and siding (Figure 7.7b), and integrated into roofing shingles and other building materials. When deposited on flexible substrates, the thin films can be rolled up and easily transported for use in remote applications. Calculators, watches, and other portable electronic equipment predominantly make use of amorphous technology. Of the three types of silicon cells, amorphous cells are the least expensive to manufacture but also have the shortest lifespans. Without the stability of a crystalline structure,

FIGURE 7.7a Lacking a crystal structure, amorphous silicon can be applied in an extremely thin film, making it flexible. Amorphous silicon is not as stable, however, and its performance degrades more quickly than cells made of crystalline silicon. *Source:* United Solar Ovonic/NREL.

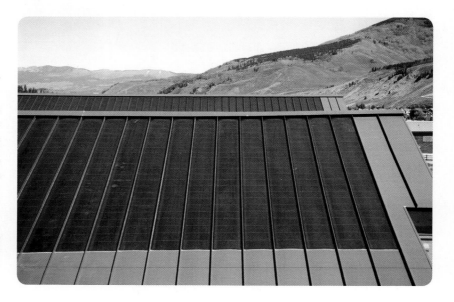

FIGURE 7.7b Unlike modules made of crystalline silicon, thin-film amorphous silicon can be adhered directly to metal roofing, avoiding the need to mount rigid frames that may sometimes compromise structure and aesthetics.
Source: Jim Yost/NREL.

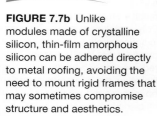

amorphous cells degrade much more quickly than monocrystalline or polycrystalline cells, losing as much as 5% of their rated power output per year.

Limitations on Efficiency

Although silicon cells currently dominate the photovoltaic industry, they are subject to inefficiencies that limit their ability to convert sunlight into electricity. Most of the solar energy that strikes a silicon cell is lost before it can be converted, and the maximum theoretical efficiency of a single silicon cell is only 29%. The greatest energy loss occurs because only a small portion of the solar spectrum participates in the photovoltaic effect. More than half of the energy in sunlight has wavelengths too long to excite electrons in silicon. These longer wavelengths simply pass through the crystal or are converted into waste heat. Other wavelengths are too short and deliver too much energy. The extra energy of these wavelengths is also wasted as heat. Some of the light reflects from the surface despite the antireflective texturing and coating, and some is blocked by the thin metal strips. Furthermore, some of the electrons that do get excited relax back into their ground states before they can migrate to the surface and pass through an external circuit.

The ideal photovoltaic material would absorb the entire solar spectrum rather than a narrow range of wavelengths. While this may never be fully realized, laminating numerous layers of different thin-film materials has increased overall conversion efficiency to more than 40%. Even more promising but still on the frontiers of research is nanotechnology, which blends particles on the scale of a billionth of a meter with semiconducting polymers. By optimizing the distribution of particle sizes, nanotechnology may be able to create thin-film materials that absorb most of the solar spectrum.

Techniques to boost output and increase efficiency include using reflective surfaces to focus sunlight on photovoltaic materials and designing internal reflections to trap light. While these techniques can improve performance, most solar cells become less efficient as their temperature increases, limiting the usefulness of concentration and trapping techniques without simultaneously improving the ability of the photovoltaic material to absorb the energy.

Nonsilicon Solar Cells

Numerous materials other than silicon are suitable for creating photovoltaic cells. Non-silicon thin-film semiconductors include those made of copper indium gallium diselenide and cadmium telluride, both of which provide greater efficiency than thin-film amorphous silicon, though their manufacture includes toxic and potentially carcinogenic compounds. Experimenters are also applying these types of thin films using ink-jet printers, which could simplify production of photovoltaic cells and reduce costs.

Organic polymers can also be made into photovoltaic material, producing solar cells that are lightweight, flexible, and potentially less costly than silicon. Currently, these cells are less efficient and undergo greater degradation than the silicon cells.

Light-absorbing dye is a technique that mimics part of the natural process occurring in green plants through photosynthesis. A dye absorbs sunlight and creates free electrons that can be harnessed in an external circuit. Although still in the experimental stages, its relatively simple manufacturing process and its use of common materials make it attractive as a possible low-cost alternative to silicon.

Figure 7.8 shows a comparison of various cell efficiencies.

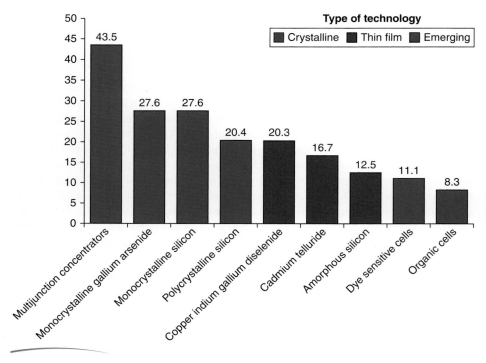

FIGURE 7.8 The maximum solar cell efficiencies recorded in meticulously controlled laboratory settings establish benchmarks for commercially available cells. The most efficient cells, multijunction concentrators, laminate three different semiconducting materials, each absorbing a different portion of the solar spectrum, and concentrate sunlight using lenses or reflectors. Monocrystalline cells made of silicon and gallium arsenide perform similarly, achieving a laboratory efficiency of almost 28%, more than twice the best efficiency of amorphous silicon. © 2016 Cengage Learning®. *Source:* NREL (2011).

Photovoltaic Electricity

All photovoltaic cells made of silicon produce a voltage of about 0.5 volt per cell regardless of cell size. This voltage is determined by the characteristics of silicon and the doped phosphorous and boron atoms. The amount of current generated by silicon cells, on the other hand, directly depends on the size of the cell and the intensity of the arriving sunlight: the larger the cell and the brighter the light, the greater the current. Since electrical power is the product of voltage and current, a greater current generates more power. Current produced by a photovoltaic cell is in the form of direct current (DC) electricity, which typically must be converted to standard household alternating current (AC) electricity before powering common appliances.

Individual photovoltaic cells are assembled into complete **photovoltaic modules** and protected behind tempered glass. Photovoltaic modules usually consist of 36–72 individual cells, each connected in series to its neighbors, to create a total voltage of about 18–36 volts. Modules are then wired together to create a **photovoltaic array**, as shown in Figure 7.9.

Series/Parallel Arrangement

Although the individual cells comprising a module are wired in series, the modules that comprise an array can be wired together either in series or in parallel.

- **Series**: When wiring modules in series, the *voltages* of the individual modules add together but the current remains the same as that of a single module.
- **Parallel**: When wiring modules in parallel, the *currents* of the individual modules add together but the voltage remains the same as that of a single module.

Most small arrays wire modules together in series to produce greater voltage and maintain low current, since larger currents require larger wires, which increase the cost of the system. However, many electrical components cannot safely exceed a voltage of 600 volts, so larger arrays often employ two or more series strings in parallel with each other. The total power output is independent of the arrangement, since power is the product of voltage and current.

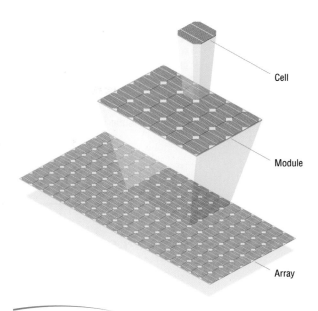

FIGURE 7.9 A solar array is the total sunlight-collecting surface and is usually made of numerous modules. Each module is comprised of many individual solar cells.
Source: U.S. Department of Energy.

Rated Power Output

The rated power of a photovoltaic module (measured in watts) represents the output achieved under standardized test conditions. The **standard test conditions** used around the world assume a solar intensity of 1000 watts per square meter (317 $Btu/ft^2/h$) at a temperature of 25°C (77°F) with no wind. This condition approximates full sunlight at sea level with the module facing directly toward the sun.

EXAMPLE 7.1

A photovoltaic array consists of 12 modules, each with an area of 1.0 m^2. If the modules have an overall efficiency of 18%, determine the power produced by the array during standard test conditions.

Solution:

The total area of the array is 12 × 1.0 m^2 = 12 m^2. The amount of solar energy striking the array under standard conditions is 1000 watts/m^2 × 12 m^2 = 12,000 watts. If the modules are 18% efficient at converting sunlight into electricity, the electrical output of the array is 0.18 × 12,000 watts = 2160 watts of electricity. This is enough to operate a typical microwave oven and hair dryer simultaneously.

EXAMPLE 7.2

A photovoltaic module with 72 individual cells is rated to produce 180 watts of power under standard test conditions. If the module operates at 36 volts, what current does the module produce under standard test conditions?

Solution:

Power (P) is the product of current (I) and voltage (V) (P = V × I). Therefore:

$$I = P/V$$

$$I = 180 \text{ watts}/36 \text{ volts}$$

$$I = 5.0 \text{ amps}$$

EXAMPLE 7.3

A photovoltaic array consists of 30 modules, each producing 180 watts at 36 volts and 5.0 amps under standard test conditions. The installer decides to create two series strings of 15 modules and combine the two strings in parallel. Determine the output voltage, current, and total power of the array at standard test conditions.

Solution:

Each string of 15 modules connected in series will produce a total voltage of 15 × 36 volts = 540 volts. The current of the string remains at 5.0 amps.

When these two strings are combined in parallel, the voltage remains the same at 540 volts, but the current increases: 2 strings × 5.0 amps/string = 10 amps.

The output of the array is, therefore, 540 volts at 10 amps, which produces a total power of 540 volts × 10 amps = 5400 watts (5.4 kW).

Photovoltaic Systems

Photovoltaic arrays produce electricity anywhere the sun is shining. In regions where there is no electric utility grid, photovoltaic arrays offer valuable electricity that may otherwise be nearly impossible to obtain. For some applications, such as operating fans and water pumps, photovoltaic arrays can be wired directly to the load (Figure 7.10). This is the simplest of all photovoltaic systems.

For most other applications, however, direct wiring is inadequate because sufficient daily variability occurs in the sun's intensity to produce fluctuations in power that would damage electrical equipment. Two methods help smooth the naturally variable output of photovoltaic arrays: (1) using the array to charge a battery bank and (2) integrating the array with the utility grid.

Isolated (Off-Grid) Photovoltaic Systems

A photovoltaic system that is off-grid (independent of the utility grid) requires a battery bank like the one shown in Figure 7.11 to provide steady power. All electrical equipment

FIGURE 7.10 Comprising the simplest of all photovoltaic systems, these modules directly power a DC pump, moving water only when the sun is shining. *Source:* Jerry Anderson/NREL.

FIGURE 7.11 A battery bank stores energy from the solar array and provides stable electricity to household appliances. Batteries must be checked and maintained regularly, enclosed to prevent accidental contact, and vented to the outside. Courtesy of Scott Grinnell.

operates from the battery bank, not the array. The photovoltaic array simply maintains adequate charge on the batteries by adding energy whenever the sun shines. A bank of batteries not only smooths the variability inherent in sunlight but also allows appliances to draw peak loads that otherwise exceed the capacity of the photovoltaic system. Unfortunately, batteries are expensive, require regular maintenance, and can be hazardous if improperly handled. Furthermore, batteries must be stored in ventilated spaces and last only 7–9 years on average. After their useful life, batteries must be recycled properly to prevent toxic compounds, such as lead, from entering the environment and posing human health concerns.

Batteries store DC electricity. Since most household appliances require AC electricity, an **inverter** must be added to the system to convert the DC electricity of the battery into standard household AC power. Inverters make this conversion with 90%–98% efficiency, offering a practical method of utilizing modern appliances when isolated from the grid.

Grid-Connected Photovoltaic Systems

When a photovoltaic array is connected to the utility grid, the grid behaves like an enormous battery. When the array provides more power than is used on site, the grid conveys the extra electricity to other customers down the line. When the array is inadequate to satisfy the energy demands on site (such as at night or during cloudy weather), the grid makes up the difference. **Net metering** records the difference between the excess solar electricity delivered to the grid and the utility-produced electricity drawn from the grid. Utility companies that offer net metering charge the customer only for the net consumption and provide payment if the photovoltaic array produces more power than is used.

In order for a photovoltaic array to interact with the utility grid, the direct current produced by the array must pass through a **synchronous inverter**, which transforms the electricity into alternating current and synchronizes the voltage and frequency to match those of the grid (Figure 7.12).

Because the electricity produced by the photovoltaic array is always synchronized with the utility grid, the inverter shuts down production from the array whenever a power outage occurs on the grid, even if sunlight is available that could otherwise provide power to the site. Consequently, some grid-connected systems also integrate a battery bank into the system, which allows the system to automatically switch over to an off-grid arrangement and maintain power during a utility outage. Batteries can, therefore, be useful for either off-grid or grid-connected arrays.

Mounting Options for Photovoltaic Arrays

Photovoltaic modules must face directly toward the sun to maximize power production. The flexible nature of wires provides this freedom, enabling photovoltaic arrays to track the sun. This is a major advantage over solar hot water systems, which are constrained by rigid piping.

The sun moves in two directions: its compass direction (rising in the east and setting in the west) and its elevation in the sky (which is a maximum at noon).

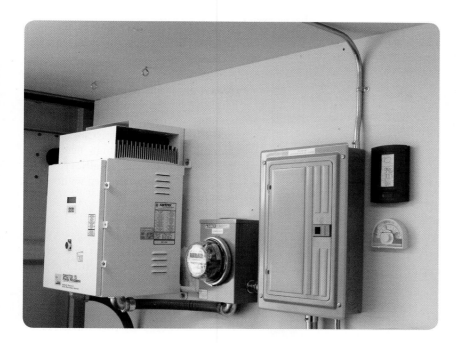

FIGURE 7.12 A synchronous inverter (the white box) takes DC power from the solar array and converts it to standard AC power, matching the voltage and frequency of the utility grid. A meter records the energy created before feeding it into the household service panel. The service panel distributes the electricity to local loads and sends any surplus to the utility grid for other customers down the line. *Source:* Abigail Krich/NREL.

Dual-axis tracking accounts for both of these directions, while single-axis tracking follows only the daily east-to-west movement. Compared to a fixed array that is manually adjusted each season to face the sun, dual-axis trackers can produce 30%–35% more energy over the course of the year. Single-axis trackers achieve an additional 20%–25% over fixed arrays. This extra energy occurs primarily during the summer, when the sun makes a higher and broader sweep across the sky.

Fixed Mount

While tracking the sun is optimal for maximum energy production, it is not always practical due to mounting options, shading conditions, integration into building components, aesthetics, and so forth. Fixed-mount systems have several noteworthy advantages:

- They provide the greatest diversity of mounting options: on the ground, against a wall, atop a roof, or attached to a support pole (see Figure 7.13).
- They can be mounted on adjustable racks so that their tilt angle can be manually repositioned each season to accommodate the sun's changing altitude.
- They permit building-integrated systems such as photovoltaic roofing shingles or thin films applied to roofing, siding, or windows. Building-integrated systems can be more cost effective and less aesthetically disruptive than stand-alone systems.
- They do not have moving parts, avoiding the need for lubrication, routine maintenance, and replacement of worn gears.

FIGURE 7.13a Ground-mounted arrays greatly simplify routine maintenance, such as removing snow, and are often preferred over roof-mounted arrays in snowy climates. The flat plate solar hot water collector, a portion of which is visible above the photovoltaic array, will be more difficult to clean. © Robbie George/National Geographic/Getty Images.

- They offer nearly the same energy production for sites that experience morning and evening shading.
- They are often the best choice for off-grid applications. When off-grid, excess summer energy produced by trackers cannot normally be utilized anyway, and a photovoltaic array sized to accommodate the building's needs during the winter will typically provide a surplus in the summer even without the benefit of a tracker.

Tracking Arrays

Tracking the sun maximizes the array's power output and is most useful for grid-tied systems with clear solar access from sunrise to sunset. Tracking arrays, such as those shown in Figure 7.14, optimize the collection of energy in two ways: (1) by exposing the greatest possible area to the sun's energy, and (2) by reducing reflection caused by oblique illumination. Usually the grid or a battery bank powers the tracking apparatus, not the sun, which permits the array to reposition itself even in the absence of sunlight (such as after sunset in preparation for sunrise).

Steel poles anchored in foundations of concrete typically support tracking arrays. The mounting pole must be tall enough for the array to tilt vertically to capture the morning and evening light, which usually requires at least 8 feet above the ground with another 4–6 feet buried. The use of steel and concrete, as well as the tracking unit itself, adds to the embodied energy of the system and the overall cost.

Most trackers can be operated manually and locked in a fixed position. This can be advantageous for snow removal or to protect the array from hailstorms and severe wind, a feature unavailable with fixed arrays.

A potential problem with trackers is that they require maintenance, can be damaged by ice accumulation, hampered by birds' nests and insect infestations, and present a possibility for failure in an otherwise extremely reliable technology.

FIGURE 7.13b For climates that do not experience significant snow, south-facing roofs offer practical mounting options, particularly when yard space is limited. *Source:* groSolarÂ©/NREL.

FIGURE 7.14 For grid-tied applications in wide-open sites, dual-axis tracking systems maximize power production. These three arrays each utilize eight polycrystalline modules.
Source: Warren Gretz/NREL.

Grid-Tied System for Residential House

Location: Northern Wisconsin (46.5° N latitude)

Mounting System	Module Type	Array Size	PV Power Rating	Date of Installation
Pole-mounted dual-axis tracking	Monocrystalline silicon	12 modules of 175 watts each	2.1 kW	2008

Description of System: Twelve monocrystalline modules, each rated at 175 watts, comprise a 2.1-kW dual-axis tracking array. Each module produces a voltage of 35.4 volts and a current of 4.95 amps under standard test conditions. The installers wired the modules together as two series strings of six modules each, creating a total operating voltage of 212 volts for each string and a total current of 9.9 amps.

A 16-foot-long steel mast, buried 6 feet in the ground and embedded in concrete, supports the dual-axis tracking unit. A sturdy aluminum rack system attached to the tracker bears the weight of the 12 modules. Even on very windy days, the tracker and aluminum rack keep the modules stable and optimally oriented.

Slight morning and evening shading blocks roughly 8% of the yearly solar resource, a small but not insignificant reduction.

Wires carrying DC current from the solar array travel 65 feet through an underground conduit to the residence, where a disconnect switch allows the solar array to be isolated from the building. The DC current then passes through a synchronous inverter, which converts the electricity from 212 volts DC to 240 volts AC. A second disconnect switch allows the inverter to be isolated from the grid. The electricity passes through the service panel, where it is distributed to household loads or continues out onto the utility grid.

Cost Analysis: At the time of installation, monocrystalline modules sold for just over $5 per watt, and the complete system cost almost $10 per watt.

System Component	Cost
(12) 175-watt monocrystalline modules (Sharp Corp.)	$10,680
Dual-axis tracker unit (Wattsun AZ-225)	$4200
Pole mount (labor and materials)	$1928
Synchronous inverter (Fronius IG 2000)	$1918
Wiring, circuit breakers, fuses, conduit, etc.	$1134
Shipping	$400
Total Cost	$20,260
State incentive program (Focus on Energy)	−$6552
Federal tax rebate	−$6078
Net Cost	$7630

An incentive program for renewable energy offered by the state of Wisconsin, called Focus on Energy, paid for 30% of the installation cost. A federal tax rebate provided another 30%, reducing the net cost to only 40% of the actual installation cost. Programs such as these encourage emerging technologies in situations where they may otherwise be unaffordable.

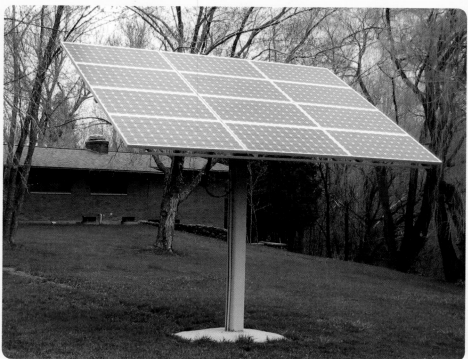

Courtesy of Scott Grinnell.

The electrical energy produced by the tracking photovoltaic array is roughly 3400 kWh/year, about 87% of the annual household demand of 3900 kWh. The local utility company pays an average of $0.11/kWh for the solar electricity, saving the customer $0.11/kWh × 3400 kWh/year = $374/year in electric bills. If the price of electricity remained constant, the time required to pay off the investment would be $7630/($374/year) = 20 years. This reflects a 5% return on investment.

Sizing a Photovoltaic System

Sizing a photovoltaic system to meet the energy needs of a site depends on a variety of factors including the solar resource, the type of mounting (fixed or tracking), whether the system will be isolated from or integrated with the utility grid, and the conservation habits of the residents.

Solar Resource

A map of the solar resource for the United States, shown in Figure 7.15, indicates the energy that would be received each day on average by a fixed array tilted at the location's latitude. A tracking array can gather up to 30%–35% more energy on average than the values indicated.

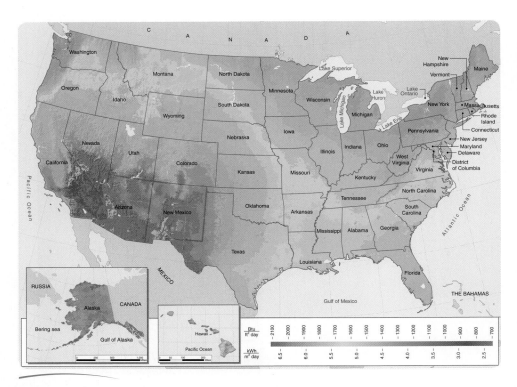

FIGURE 7.15 The solar resource for the United States varies from less than 3 kWh/m²/day in parts of Alaska to almost 7 kWh/m²/day in the Southwest. Unlike solar hot water systems that require bright illumination, photovoltaic systems produce some electricity even on cloudy days, making them useful even in regions with low solar resource. © 2016 Cengage Learning®. *Source:* Adapted from the National Renewable Energy Laboratory (NREL).

The solar resource indicated in Figure 7.15 assumes the absence of local shading. The presence of nearby trees, buildings, poles, wires, and so forth can reduce the actual solar resource considerably. Completing a site assessment that measures shading is especially important before installing photovoltaic systems because photovoltaic modules are more sensitive to shading than solar hot water collectors. Every solar cell within a module is connected in series with its neighbors, requiring current to flow through every cell. A shaded cell acts as an impediment to its neighbors and decreases the output of the entire module. Hence, even partially shading a module can severely reduce its output. On the other hand, unlike most residential solar hot water systems, photovoltaic arrays continue to collect energy even on overcast days, though at a reduced level.

Grid-Tied Systems

For grid-tied systems, sizing an array to generate all of a site's net energy demands requires dividing the total annual electricity consumption (in kWh) by the annual solar resource, which equals the value provided in Figure 7.15 (shown in kWh/m²/day) multiplied by 365 days per year. Multiplying this product by the standard test condition of 1 kW/m² provides the **PV power rating** of the array, which is generally the power output of modules advertised by retailers.

In practice, however, photovoltaic systems suffer numerous losses that reduce the actual electricity produced. These losses include dust on the modules, resistance in wires and connectors, inefficiencies associated with inverters and/or batteries, and reduced cell performance due to excessive temperature. Together these amount to as much as 20% or more of the potential power output. Including this factor increases the size of the array by a factor of 1/0.80 or 1.25.

$$\text{PV power rating} = 1.25 \left[\frac{\text{Annual energy consumption (kWh/year)}}{\text{Solar resource(kWh/m}^2\text{/day)} \times 365 \text{ days/year}} \right] \times 1 \text{ kW/m}^2$$

EXAMPLE 7.4

A grid-connected house in San Luis Obispo, California, uses 5800 kWh/year of electricity. The owners would like to install a fixed photovoltaic array capable of generating their annual usage. If the solar resource for San Luis Obispo is 6.3 kWh/m²/day, determine the necessary power rating of the array. Assume average system losses of 20%.

Solution:

$$\text{PV power rating} = 1.25 \left[\frac{5800 \text{ kWh/year}}{6.3 \text{ kWh/m}^2\text{/day} \times 365 \text{ days/year}} \right] \times 1 \text{ kW/m}^2$$

$$\text{PV power rating} = 3.2 \text{ kW}$$

This could be done with 18 180-watt modules and would cover an area of 250 square feet.

CASE STUDY 7.2

Grid-Tied System for Net Zero Energy Residence

Location: Northern New York (42.4° N latitude)

Mounting System	Array Specifications	Photovoltaic Power Rating (total)	Date of Installation
Fixed mount on roof	28 215-watt mono-crystalline modules	6.0 kW	2011

Description of System: In 1998, the owners purchased a spacious two-story church overlooking the Hudson River in upstate New York. Renovating the church created a 5200-square-foot home and office spaces through meticulous restoration of original architecture and complete upgrades of electrical and plumbing systems. Additional insulation in the walls and floors and energy-efficient appliances reduce heat loss and electricity consumption. A propane in-floor hydronic heating system, augmented by a wood-burning stove, maintains a comfortable year-round temperature.

In 2011, the owners installed 28 215-watt monocrystalline photovoltaic modules on the south-facing asphalt shingle roof of the church. A flush-mount system anchored to the roof tilts the modules at roughly 30°. The relatively shallow pitch of the modules enhances summer production, when fewer days of cloud cover obstruct sunlight. Since the array is grid connected, the high summer production more than compensates for the reduced winter production, resulting in more total energy generation than a system mounted at the theoretically optimal angle of 42° (the location's latitude).

Cost Analysis:

Description	Cost
Gross system cost	$38,120
New York State incentive program ($1.75 per watt)	−$10,535
Out-of-Pocket Cost	$27,585
Federal income tax credit (30% of out-of-pocket cost)	−$8276
New York income tax credit (25% up to $5000 maximum)	−$5000
Final Customer Cost after Incentives and Credits	$14,309

The cost of the 6.0 kW system, including modules, mounting racks, wiring, inverter, and monitoring system, totaled about $14,300.

The array produces about 7000 kWh per year of electricity, providing the owners with an annual savings of over $1000. This represents a return on investment of over 7%, and the system will pay for itself in about 14 years, assuming the cost of electricity remains constant over time.

Courtesy of Scott Grinnell.

Off-Grid Systems

For most systems independent of the utility grid, no meter keeps track of the annual electricity consumption. If this value is unknown, it must be calculated by assessing everything on site that consumes electricity, a procedure called a load analysis. This requires an itemized list of every electrical device (lights; appliances; heating, ventilating, and air conditioning systems; fans; water pumps; tools for hobbies and work; rechargeable batteries; electronics; and so forth) and the average length of time each device will be operating during a typical day. A plug-in power meter, such as the one shown in Figure 7.16, can help determine the power consumption of various devices. Table 7.1 shows the typical power consumption of common household items.

After determining the average annual energy consumption of an off-grid site, the appropriate PV power rating for an array can be found, using the same equation presented above. However, unlike grid-tied systems, off-grid sites cannot draw extra power from the grid during prolonged cloudy weather or for occasional elevated levels of electrical demand. Therefore, off-grid systems are typically sized to be larger

FIGURE 7.16 Plug-in meters, like this one, provide information on the voltage, current, and power used by any appliance plugged into the meter, making it a useful instrument when conducting load analyses. Courtesy of Scott Grinnell.

TABLE 7.1 Average electrical use for household items.			
ITEM	**USAGE (WATTS)**	**ITEM**	**USAGE (WATTS)**
General Household		**Kitchen Appliances**	
Air conditioner	1000–1500	Blender	350
Blow dryer	1000–1200	Can opener	125–200
Ceiling fan	10–50	Coffee grinder	100
Clock radio	5	Coffee pot	800–1500
Clothes washer	500–650	Dishwasher	1200–2000
Dryer (electric)	4000–4800	Dehydrator	600
Dryer (gas)	300–400	Food processor	400
Electric blanket	175–200	Fry pan (electric)	1200
Electric clock	2–4	Garbage disposal	450

(continued)

TABLE 7.1 Average electrical use for household items. *(continued)*

ITEM	USAGE (WATTS)	ITEM	USAGE (WATTS)
General Household		**Kitchen Appliances**	
Furnace fan	300–1000	Hot plate	1200
Garage door opener	350	Microwave (0.5–1.5 ft³)	600–1500
Heat lamp (bathroom)	250	Mixer	120
Heater (electric)	1500	Popcorn popper	250
Iron	1000–1500	Oven (conventional)	3500
Radio	10	Range (large burner)	2100
Security system	3–6	Breadmaker	430
Sewing machine	100	Range (small burner)	1250–1600
Shaver	10–15	Range: Igniter glow bar of gas oven	300–400
Table fan	10–15	Refrigerator/freezer (standard, 16–22 ft³, 13–14 hr/day)	475–540
Vacuum cleaner (handheld)	100	Toaster	700–1500
Vacuum cleaner (upright)	200–960	Freezer (Energy Star, 17 ft³)	460
Water heater (electric)	4500	Toaster oven	1150
Water pump	1000	Waffle iron	800–1200
Tools		**Entertainment**	
Air compressor (1 hp)	1000	CD player	35
Band saw	1100	Cell phone (charging)	24
Belt sander	1000	Computer printer (ink-jet)	25
Chain saw (12")	1100	Computer printer (laser)	1000
Circular saw (7 1/4")	900	Computer	40–150
Disc sander (9")	1200	DVD player	25
Drill (1/2")	750	Laptop computer	20–50
Hedge trimmer	450	Satellite system	12–45
Lawn mower (electric)	1200–1500	Stereo	15
Weed eater	500	TV (19" color)	60–100
		TV (25" color)	125–150
		TV (big screen, plasma, LCD)	300

than grid-tied systems. Alternatively, owners of off-grid systems may select to include backup generators or other forms of renewable energy to supplement electrical demands when solar energy is insufficient.

EXAMPLE 7.5

The owner of a remote cabin in West Virginia, where the solar resource is 4.2 kWh/m²/day, determines the following information by conducting a load analysis. What size fixed array will satisfy the electrical demands on average?

Load	Power Consumption (watts)	Average Daily Usage (hours)	Average Daily Energy (watt-hours)
4 CFL bulbs, 23 watts each	92	4	368
Refrigerator	240	14	3360
Microwave oven	1200	0.5	600
Computer	105	6	630
Battery charger	200	2	400
Ceiling fan	20	3	60
Electric glow bar on propane oven	300	0.5	150
Clock	3	24	72
		TOTAL	5640

Solution:

On average, the cabin uses about 5.64 kWh of electricity per day, which amounts to an annual consumption of roughly 5.64 kWh/day × 365 days = 2060 kWh. A photovoltaic array that would satisfy the average demands is given by:

$$\text{PV power rating} = 1.25 \left[\frac{2060 \text{ kWh/year}}{4.2 \text{ kWh/m}^2/\text{day} \times 365 \text{ days/year}} \right] \times 1 \text{ kW/m}^2$$

$$\text{PV power rating} = 1.7 \text{ kW}$$

This could be done with 10 180-watt panels, which would cover an area of roughly 140 square feet. The cabin owner would need to include a sufficiently large battery bank to store energy for periods of darkness or extended cloud cover and may also want a backup source, such as a generator.

CASE STUDY 7.3

Off-Grid System for Homestead

Location: Northern Wisconsin (46.5° N latitude)

Mounting System	Array Specifications	PV Power Rating (total)	Date of Installation
Fixed mount on roof	12 65-watt polycrystalline modules 4 50-watt monocrystalline modules	950 watts	1999

Description of System: An 1850-square-foot house was designed and built by the owners as part of a 180-acre homestead. Located a mile from the nearest utility line, the owners opted to remain off-grid and installed a 950-watt roof-mounted photovoltaic system using a mixture of polycrystalline and monocrystalline silicon modules. Twice a year the owners adjust the modules to optimize the seasonal tilt angle.

The solar array charges a battery bank of 10 Deka® L16 lead-acid batteries (each weighing over 100 pounds), providing a usable storage of 11 kWh of electrical energy. The original battery bank provided continuous service for 9 years before being replaced. The owners monitor the battery bank regularly to verify its state of charge and perform routine maintenance, including adding distilled water each month and equalizing the batteries every other month (equalization is a process by which the batteries are intentionally overcharged for a short period of time, a procedure that extends battery life by preventing sulfate deposition and electrolyte stratification). The battery bank is housed in a protective box that is vented to the exterior and maintained at a relatively constant temperature.

The house utilizes both 12-volt DC circuits and 120-volt AC circuits, establishing ample options for lighting and appliances. Two Trace® 2500 inverters convert the 12-volt DC energy of the batteries into 120-volt AC electricity. The two inverters allow for large start-up loads characteristic of power tools with induction motors.

Four wood stoves located in different rooms throughout the house provide heat. A specially selected propane cook range minimizes electricity consumption by operating without an electric ignition bar (the ignition bar common to most gas ranges draws 300–400 watts of electricity continuously, a consumption that would unnecessarily drain the limited battery bank). In addition to selecting a special cook range, the owners installed a high-efficiency refrigerator that excludes automatic defrost cycles, a convenient but energy-intensive process included in most commercially available refrigerators.

Courtesy of Scott Grinnell.

Living off-grid requires an intentional lifestyle—an ability to plan electricity usage according to the available resource. It requires eliminating all phantom loads, acquiring high-efficiency appliances, and being constantly vigilant of demands in energy and changes in weather.

Since the original installation of the modules in 1999, the owners have observed very minimal decrease in power output, perhaps 5% overall (about 0.4% per year average).

Cost Analysis: The cost of the 950-watt photovoltaic system totaled $13,100. This includes the modules, wiring, battery bank, inverters, and charge controller. The cost of bringing electricity to the site from the utility grid, located about a mile away, would have been $25,000 in 1999.

The array produces about 550 kWh of usable electricity per year, saving the owners about $450 in annual utility costs, including the connection fees. During extended cloudy periods between November and March, a 3000-watt gasoline generator provides backup power. The generator cost $1800 and consumes about 120 gallons of gasoline per year. If the generator converts gasoline into electricity with an efficiency of roughly 20%, the 120 gallons generate 880 kWh of electricity at an annual expense of about $450/year.

The cost to replace the current battery bank was $2150 in 2010 and represents an ongoing expense every 7 to 10 years.

Element	Cost
950-watt complete system	$13,100
Backup generator	$1800
Generator fuel	$450/year
Battery bank (average annual expense)	$240/year
Total Cost	$14,900 plus $690/year
Cost to bring in utility grid	$25,000
Annual cost of electricity from grid	$560/year

Conservation

The most economical method of meeting electrical demand is to reduce consumption and conserve energy. Rather than installing sufficient photovoltaic modules to meet the current load, a more cost-effective strategy is usually to replace inefficient appliances like old refrigerators and furnaces with more efficient models and to adopt energy-conserving strategies. For example, unplugging electronics when not in use, such as computers and televisions, eliminates phantom loads that can draw enough electricity to require the installation of extra modules. Furthermore, the addition of insulation to poorly insulated buildings will pay for itself far more quickly than adding more modules to operate air conditioning or heating units. Replacing incandescent lightbulbs with compact fluorescent (CFL) or light-emitting diode (LED) lightbulbs and turning off lights when not in use are among the first energy-saving conversions individuals can make. Professional energy audits also help determine avenues for saving energy.

CASE STUDY 7.4

Conservation

Location: Central Colorado (38.9° N latitude)

Mounting System	Array Specifications	PV Power Rating (total)	Date of Installation
Adjustable pole mount	18 75-watt mono-crystalline modules 2 85-watt monocrys-talline modules	1520 watts	2001

The rammed-earth tire structure described in Chapter 4 (*Case Study 4.1: Rammed-Earth Tire House*) demonstrates the role of conservation in reducing the size (and, therefore, the cost) of a photovoltaic system. Located in the mountains of Colorado at a site that experiences cold but sunny winters, the owners designed a comfortable home utilizing a relatively modest 1.52-kW photovoltaic array for a quasi-off-grid system.

The system includes a 24-volt battery bank capable of storing 1160 amp hours of energy—equivalent to a little more than 27 kWh. To preserve the life of the batteries, however, only roughly half this amount of energy is ever withdrawn at one time before recharging the batteries. To accommodate occasional peak demands of energy or to buffer rare periods of prolonged cloudy weather, the owners chose to become grid connected. However, this connection is rarely utilized. At the time of installation (in 2001), the local utility company did not offer the option of net metering. While connecting with the grid eliminated the need for supplemental systems (such as a gasoline generator), it did not permit selling surplus electricity back to the utility. The owners, therefore, sized the photovoltaic array as if the system was isolated and used the grid connection primarily as insurance against unpredictable weather. A well-designed passive solar home with energy-efficient appliances and conservative habits allow the residents to use a mere 5 kWh/day of electricity on average—less than 2000 kWh/year. Compact fluorescent lighting and Energy Star appliances reduce electrical demands, and the owners conscientiously eliminate phantom loads by unplugging other electrical equipment when not in use. All total, the utility grid supplies a mere 100 kWh/year, a small fraction of the typical 8000 kWh/year consumed by a conventional home in the same area.

An adjustable pole-mounted system supports 20 monocrystalline photovoltaic modules and permits the owners to change the tilt angle seasonally to optimize performance. The photovoltaic array, battery bank, AC inverter, and controls cost a total of $16,000 in 2001.

While the owners could have installed a larger system, they instead found ways to use less energy, while maintaining the same level of comfort. This not only saved on installation costs but also lessened the embodied energy and resource depletion associated with additional modules and batteries.

Courtesy of Jerry D. Unruh and Diana P. Unruh.

Solar Thermal Electricity

Solar thermal technology is another method of converting the sun's abundant energy into electricity. Unlike the direct conversion that occurs in photovoltaic modules, solar thermal power plants operate very similarly to conventional power plants, except they burn no fossil fuels, require no smokestacks, and generate no pollution. Solar thermal power plants can even store thermal energy using molten salts and generate solar thermal electricity when the sun is not shining. However, like conventional power plants, solar thermal plants require cooling water and cannot easily be deployed in deserts, where water is scarce.

The two most common methods of concentrating the sun's rays to generate electricity employ a parabolic trough system or a solar tower.

Parabolic Trough

A parabolic trough power plant collects solar energy in long lines of parabolic mirrors that track the sun's east-to-west movement. The mirrors concentrate sunlight by a factor of 80 times or more onto metal absorber pipes embedded in evacuated tubes. The evacuated tubes nearly eliminate heat loss by conduction and convection, and a selective coating on the absorber pipes minimizes heat loss by radiation. These features allow the system to heat oil to temperatures as high as 750°F (400°C). A heat exchanger transfers energy from the hot oil to water, which boils into steam and spins turbines connected to an electrical generator. Cooling towers condense the steam back into water and continue the cycle. (See Figure 7.17.)

FIGURE 7.17a Highly reflective parabolic troughs focus the sun's energy onto fluid-filled absorber pipes, creating very high temperatures that can produce steam and drive a turbine. Unlike photovoltaic arrays, these systems can store thermal energy for use at night, which allows them to produce electricity uniformly throughout the day. *Source:* SkyFuel, Inc./NREL.

FIGURE 7.17b A commercial power plant may consist of hundreds of parabolic troughs, each capable of tracking the sun's east-to-west motion. The heated fluid passes through a heat exchanger to boil water into steam, which drives a conventional turbine. A cooling tower then condenses the steam back into water so that the cycle can continue. This power plant, built in 1990 at Kramer Junction, California, produced as much as 30 million watts of electricity. *Source:* Sandia National Laboratories/NREL.

Solar Tower

A **solar tower power plant** consists of hundreds of pole-mounted dual-axis tracking mirrors that direct sunlight onto a receiving chamber in a central tower (Figure 7.18). Each mirror is computer controlled to ensure that sunlight is focused directly on the receiving chamber, which can reach temperatures of 1800°F (1000°C). Hot air or molten salts within the receiving chamber store and transfer this energy to water to create steam and drive electrical generators.

FIGURE 7.18 First constructed in 1982, this experimental power plant in Barstow, California, produced as much as 10 million watts of power by reflecting sunlight with hundreds of dual-axis tracking mirrors onto a receiving chamber at the top of a central tower. Although no longer in use, it operated successfully for numerous years. *Source:* Sandia National Laboratories/NREL.

Chapter Summary

Sunlight can create electricity directly through the photovoltaic effect or by boiling water to operate a steam turbine. Although heating water to generate electricity can be economical at commercial scales, most residential solar electricity comes from photovoltaic arrays. Photovoltaic arrays can be as small as a single module to operate a remote communication signal or as large as a commercial power plant that utilizes thousands of modules integrated with the utility grid. The ability to create electricity at almost any scale endows photovoltaic technology with a versatility that makes it widely applicable.

Photovoltaic materials convert solar energy into electricity by directing electrons that have been excited by sunlight through an external circuit before they relax back to their natural state. Two thin layers of optimally selected semiconducting materials—usually composed of silicon doped with phosphorous and boron—comprise most of the photovoltaic cells available today.

Due to the high costs of production, generating electricity from photovoltaic modules is currently one of the most expensive forms of renewable energy. However, advancements with thin-film technology using organic polymers, light-absorbing dyes, and nanoparticles promise to increase efficiency and reduce costs. These advances may also promote building-integrated systems and give rise to many new products.

Photovoltaic arrays provide electricity for isolated stand-alone systems, as well as systems integrated with the utility grid. Photovoltaic arrays can be mounted nearly anywhere, from rooftops to automobile tops, from the sundecks of cruise ships to the wings of airplanes, from ground mounts to top-of-pole trackers. They can be protected behind tempered glass in rigid panels or applied as peel-and-stick films to roofing and siding. The versatility of photovoltaic materials makes them highly adaptable to society's rapidly changing technological needs.

Review Questions

1. Photovoltaic modules made of monocrystalline silicon cells degrade slowly with time. Suppose this rate is equal to 0.5% per year, so that after 10 years a module can be expected to operate at an efficiency of only 95% of its original output (5% less). What would you consider to be the useful lifespan of a solar module made of monocrystalline cells?

2. Plans to cover large tracts of desert with photovoltaic modules have drawn criticism from those concerned over the impact on desert ecosystems. Proponents contend that producing the same amount of power from a coal-fired power plant usually disturbs even more land when including strip mines that provide most of the coal. Compose an argument for or against the deployment of large desert photovoltaic arrays.

3. When considering types of silicon modules, in what applications are monocrystalline modules most appropriate? When are amorphous modules most appropriate?

4. A solar pathfinder indicates that shading occurs at a grid-connected site before 10 A.M. and after 3 P.M. throughout the year. What kind of photovoltaic mounting system would you recommend (fixed, single-axis tracking, or dual-axis tracking)?

5. An off-grid campground located in the open prairies of South Dakota operates only during the summer months. The proprietor of the campground wishes to install a photovoltaic system. Which of the three types of solar cells would you recommend? What type of mounting would you recommend? Explain your reasoning.

6. A location in southern Texas experiences mild winters and hot summers with abundant sunshine all year, particularly during the summer. At what time of the year do you suppose a photovoltaic array would be least able to meet the electrical demands of a *conventional* building? Why?

7. Compare flat plate solar hot water collectors with silicon photovoltaic modules. Address each of the following:
   a. Efficiency at converting direct sunlight to usable energy (either hot water or electricity)
   b. Ability to harness solar energy on overcast days
   c. Expense of installing a system
   d. Mounting options

Practice Problems

1. The maximum voltage of a photovoltaic array is 600 volts. Assume a particular array will be utilizing 180-watt modules (36 volts, 5 amps).
   a. What is the maximum number of modules that can be wired in series?
   b. How would you wire together 48 modules?
   c. What would be the output voltage and current of the 48-module array?

2. A home energy audit recommends replacing 14 lightbulbs, disconnecting several phantom loads (using a power strip), and purchasing a more efficient refrigerator and chest freezer, as indicated in table at bottom of the page.
   a. Determine the annual reduction in energy consumption (in kWh) after making the indicated changes.
   b. If electricity costs $0.12/kWh, calculate the annual savings.
   c. How long will it take for the annual savings to pay off the cost of replacement?

3. Instead of reducing energy consumption by replacing appliances as recommended by the energy audit above, a homeowner in San Antonio, Texas, decides to increase the size of the solar array to accommodate the larger load. A local installation company charges $7 per watt of rated capacity for roof-mounted polycrystalline silicon arrays.
   a. What is the approximate solar resource of San Antonio, Texas?
   b. Calculate the rated capacity of a photovoltaic array that would produce the extra electricity necessary to equal the energy saved in Problem 2.

c. How much would it cost to install a system capable of providing the annual energy saved in Problem 2?
d. How long would it take to pay off the installation of a system to generate the annual energy saved in Problem 2?
e. What can you conclude about the value of conservation?

4. A grid-tied system with a rated capacity of 10 kW is able to provide all the power for a small commercial building. The average annual energy consumption of the site is 14,600 kWh/year.
   a. What is the solar resource at this site?
   b. Give two locations within the United States that offer this solar resource.

5. A solar pathfinder indicates that a fixed array in Seattle, Washington, experiences a 25% loss of sunlight due to shading. For a grid-tied building that consumes 6000 kWh/year of electricity, what size array will provide all of the electricity on average? Assume standard system losses.

6. A camel transports a chest of frozen medical supplies across a desert to a remote village in Africa. The chest consumes 80 watts of power continuously to keep the contents frozen. The solar resource across the desert is 6.8 kWh/m^2/day. What size module (in m^2) is required to power the chest both day and night? Assume a 20% loss. Convert your answer to square feet (1 m^2 = 10.76 ft^2).

7. A remote warning beacon in Lake Superior flashes a 100-watt signal 12 hours each day. Determine the rated capacity of a photovoltaic system that could provide all of the required electricity.

Item Replaced	Daily Energy Consumption of New Product	Daily Energy Consumption of Old Product	Amount of Energy Saved Each Day	Cost of Replacement
14 CFL bulbs, each 23 watts	1.3 kWh (when on 4 hours per day)	5.6 kWh (when on 4 hours per day)	4.3 kWh	$28
Phantom loads amounting to 30 watts	0	0.7 kWh	0.7 kWh	$5
Chest freezer	0.7 kWh	1.8 kWh	1.1 kWh	$350
Refrigerator	0.9 kWh	2.2 kWh	1.3 kWh	$650

Assume a solar resource of 3.9 kWh/m^2/day and a 20% loss.

8. A solar-powered garden light is designed to collect sunlight during the day with a 4-square-inch amorphous silicon solar cell. Manufacturers claim the solar cell can illuminate a small LED bulb for up to 8 hours.

    a. If the bulb draws 0.055 watt and operates for 8 hours, how much energy (in kWh) must be stored in the garden light's battery?

    b. What solar resource do the manufactures assume (in kWh/m^2/day) in order to adequately charge the garden light? Assume the solar cell is 5% efficient at converting sunlight into electricity. An area of 4 square inches equals 0.00258 square meter.

9. Satellites operating above the earth's atmosphere receive a solar intensity of about 1370 watts/m^2 and can remain in the sunlight continuously by avoiding the earth's shadow. For such a satellite, there is no night.

    a. For a system rated at 800 watts under standard test conditions, what would be its output above the earth's atmosphere?

    b. The solar resource given in Figure 7.15 accounts for a variety of earth- and atmosphere-related considerations, such as weather conditions, length of day, and angle of illumination. These don't apply to a satellite whose modules always face directly toward the sun. Determine the solar resource for such a satellite.

10. A grid-connected dual-axis tracking array mounted on a pole in a wide-open field near Barstow, California, gathers 35% more solar energy than indicated by Figure 7.15. The array is used to power a household that consumes an average of 6600 kWh per year.

    a. Estimate the solar resource for this array.

    b. Determine the photovoltaic power rating that will provide all of the average electrical demands of the home. Assume a standard loss of 20%.

11. The cabin in Example 7.5 consumes an average of 5.6 kWh of electricity per day. Suppose the owner wants to install an array such that a single sunny day will charge the battery bank with enough energy to power the cabin for a full week. Use a solar resource of 4.2 kWh/m^2/day.

    a. How much energy must the array produce during a single sunny day?

    b. Determine the photovoltaic power rating, assuming a fixed array and a 20% loss.

    c. If the system costs $7 per watt to install, what would be the total expense? Is this reasonable?

12. Suppose prolonged cloudiness during the months of November and December results in direct sunlight suitable for producing hot water from flat plate collectors only 3 days per month. Each month, 20 additional days provide diffuse light, permitting photovoltaic modules to operate at roughly 30% of their rated output. A homeowner questions whether generating electricity from the photovoltaic modules to operate an electric hot water heater might be more economical than using a flat plate collector to heat water directly. Determine which system would produce the most hot water, making the following assumptions:

    • For each of the 3 sunny days, the solar hot water system captures 60% of the available solar energy, while photovoltaic modules convert only 18% to electricity.

    • Each system operates for 4 hours per day, and the clear-day solar illumination is 1.0 kilowatt.

    • The conversion of electricity into hot water is 100% efficient.

Wind Power

CHAPTER 8

Source: Invenergy, LLC/NREL.

Introduction

Devising methods to harness the power of the wind has challenged humanity for centuries. Since the earliest civilizations, wind has been a symbol of power and inconstancy, an intermittent force that passes by and disappears, only to return again unexpectedly. Unlike the energy extracted from the sun, which is silent and subtle, the energy present in wind is exuberant and obvious, apparent to the earliest people. Generations of sailors, inventors, and strategists puzzled over methods to capture the wind's energy, crafting thousands of different designs based on varying perceptions and needs.

Although people may have used windmills in China and Babylon as early as 2000 years ago, archaeologists provide no definitive record. The ancient Greeks and Romans made no mention of using wind for commercial applications, though they described a device that used wind to drive the bellows for an organ. However, by the middle of the ninth century, the Persians relied on wind to lift water from streams for irrigation, and by the tenth century, windmills were widespread throughout the Middle East for grinding corn. The Dutch used windmills to pump water and drain lowlands, and by the sixteenth century, they had refined the windmill into the traditional form still seen today (Figure 8.1).

Early people used windmills to reduce manual labor and raise the standard of living. Where tumbling rivers were unavailable to drive mills, wind power ground grain into flour, pumped water from lowlands, and sawed logs into lumber. Windmills mashed

FIGURE 8.1 The traditional windmill common in Europe during the sixteenth century harnessed wind using cloth sails attached to four long wooden frames. The miller varied the amount of sail depending on wind conditions and removed the sails altogether during very strong winds. Today very few traditional windmills provide any profitable commercial function. © Olenandra/Shutterstock.com.

seeds for oil, pulped wood for paper, hulled cereals, and crushed flint and chalk for use in pottery and cement.

Nevertheless, the inconstancy of wind limited its usefulness, and wind power repeatedly succumbed to more dependable power sources when they became available. Furthermore, the low density and diffuse nature of air required early windmills to be large in order to develop much power. Even though energy from wind is free, these limitations always left wind vulnerable to competition from other forms of energy.

The demise of the traditional windmill began in the mid-1800s with the advent of the steam engine. The steam engine provided not only a reliable method to grind grain but also a means to ship it over great distances. The individual mills that dotted the countryside fell into disrepair as more and more materials were moved inexpensively between rural farming communities and centralized urban factories. Steam-driven locomotives, riverboats, and transoceanic merchant ships, primarily powered by coal, moved goods around the world. Coal delivered power when it was needed, not only when the wind blew, and quickly displaced the traditional windmill for most of its applications. Pumping water remained an exception.

The inhabitation of the Great Plains of the United States created a new demand for wind power (Figure 8.2). As settlers spread

FIGURE 8.2 Hundreds of thousands of multivane windmills once dotted the American Great Plains, pumping water for farms and ranches. Many continue to operate today and serve the same purpose. *Source:* Jim Green/NREL.

west, windmills provided essential water for farms and ranches, most of which were far removed from other power sources. Windmills worked well for pumping water because water could be stored in reservoirs or holding tanks that buffered the inconstancy of wind. The traditional windmill with four large sails gave way to a windmill with many short vanes, called a windpump. The shorter vanes allowed the windpump to start turning with less wind but sacrificed efficiency and power. However, because water could be pumped slowly, the multivaned windpump proved ideal. By the 1890s, the American windpump, usually mounted atop a lattice tower, had become a hallmark of human habitation throughout the Great Plains. Cowboys not only tended cattle and mended fences but also learned to grease the gears of windmills.

Efforts to produce electricity with windmills began as early as the 1880s (see Figure 8.3), but the output of windmills proved too irregular for practical

FIGURE 8.3 In 1887, the inventor Charles Brush erected one of the earliest electric wind generators. Measuring 56 feet in diameter, it included a 60-foot-long tail and generated a peak power of 12 kilowatts. It operated for 12 years, providing power for Brush's home. Courtesy of the Western Reserve Historical Society, Cleveland, Ohio.

applications. Furthermore, neither the traditional European windmill nor the American windpump could be easily adapted to generate electricity because their gears were not well suited to the variable loads imposed by electrical generators. Since existing windmills required substantial redesign of their mechanism to be useful and since wind was inherently unable to guarantee a definite amount of power at any moment, very few windmills ever produced electricity.

After World War I, engineers applied advances in airplane propeller design to wind-generated electricity. These propellers created the first wind *turbines,* which generated electricity rather than milling grain or pumping water. After many variations and refinements, wind turbines can now exceed 50% efficiency at converting wind energy into electricity, and today the three-bladed variable-pitch rotors standing atop tubular towers symbolize modern wind power.

Harnessing the Wind

Mariners discovered centuries ago that wind-driven vessels could sail faster across the wind than with it. When sailing downwind, a vessel could travel no faster than the wind itself—at that speed, the sail moves away from the wind as fast as the wind approaches it, and the wind can no longer push on the sail (Figure 8.4). When sailing sideways to the wind, however, the wind continues to push on the sail no matter how fast the vessel travels (Figure 8.5).

FIGURE 8.4 The sail of this boat captures wind like a parachute, using the force of drag to push the boat downwind. Simple sails like this are generally useful only when traveling with the wind.
© David H. Wells/Photodisc/Getty Images.

Sailing across the wind employs the wind in a different way than sailing downwind. When traveling downwind, the wind simply pushes against the sail, filling it like a parachute; when moving crosswind, the sail stretches into a curved shape that deflects air around it on both sides. The air flowing across the outward-bowing side travels faster and creates a region of lower pressure. The force caused by the difference in pressure between the two sides, called the force of **lift**, is perpendicular to the curve, as shown in Figure 8.6. The vessel's keel limits sideways motion and allows the sailboat to move forward.

Like sailing vessels, wind turbines can extract energy from the wind in different ways and fall into two basic categories:

- **Drag-based design**: Much like a boat sailing downwind, these turbines catch wind with surfaces perpendicular to the wind direction: paddles, cups, or parachutes. The force acting on these surfaces is caused by **drag**—the force of the air colliding with the surfaces. They include simple devices such as the panemone (Figure 8.7a) and modern anemometers that measure wind speed (Figure 8.7b).
- **Lift-based design**: Most modern wind turbines make use of aerodynamic lift. Like the propellers of airplanes and the rotors of helicopters, the blades of modern wind turbines are slender, tapering surfaces curved into the shape of an airfoil. Lift-based wind turbines continue to extract energy from the wind even when the blades rotate faster than the wind speed.

Throughout the ages, inventors have designed a great variety of drag-based wind machines, many of which remain conceptually interesting. Nevertheless, all drag-based designs are inherently less efficient at capturing wind energy than lift-based designs, as they are limited to a maximum theoretical efficiency of only a *quarter* of that of lift-based designs. Consequently, all serious wind turbines use precision-shaped blades to harness wind energy through lift rather than drag.

FIGURE 8.5 By utilizing the force of lift, modern sailboats can navigate across wind and travel faster than the wind itself. The force of lift acts perpendicular to the sail, requiring sailboats to include a keel to limit sideways motion. Courtesy of Scott Grinnell.

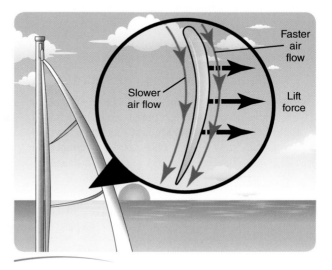

FIGURE 8.6 The curved shape of the sail causes air to move faster on the outward bowing side, reducing the pressure compared to the other side. The difference in pressure across the sail creates the force of lift. © 2016 Cengage Learning®.

(a)

(b)

FIGURE 8.7 (a) The Persian panemone—used to pump water and grind grain as early as 500 A.D.—consisted of vertical reeds fastened to lightweight wood frames that rotated about a vertical axis. The panemone represents one of the simplest but least efficient drag-based wind designs. © 2016 Cengage Learning®.
(b) The modern anemometer is a drag-based device used to measure wind speed. Cup-shaped vanes catch the wind and rotate about a vertical axis. With every rotation, the vanes swing back into the wind stream, but the drag force acting on the rounded surface is considerably less than the drag force acting on the open cup, so the vane spins away from the cupped side. Courtesy of Scott Grinnell.

Further Learning APPLICATIONS OF AERODYNAMIC LIFT

The force of lift not only permits sailing vessels to move across the wind but also allows airplanes to fly, boomerangs to return, and kites to glide in the sky. Upside-down wings, called spoilers, press race car tires to the ground for improved traction. The force of lift also enables birds to soar and fish to swim and governs the movements of bacteria, windblown seeds, and flying squirrels.

Power from Wind

The **rotor** of a wind turbine extracts kinetic energy from moving air and converts it to rotational kinetic energy of a turning shaft. This shaft drives a generator, which produces electricity. The power available to a wind turbine depends on three parameters: the wind speed, the amount of air intercepted by the rotor, and the density of air (see Figure 8.8):

- **Wind speed** (v): The wind speed determines both the kinetic energy of the air and the rate at which this energy strikes the blades of the rotor, making it by far the dominant factor in determining the power available to a wind turbine. In fact, the power of a turbine depends on the wind speed *cubed*.

- **Rotor area** (A): The amount of wind captured by a rotor is the area swept out during its rotation. This equals the area of a circle: $A = \pi r^2$, where r is the radius of the rotor (the length of each blade). Hence, the larger the rotor, the more energy it can collect.

- **Density of air** (ρ): The denser the air, the greater the impact wind will have upon striking the blades of a rotor. Cold dry air near sea level is denser and can provide more power than hot moist air at high elevations.

These three parameters create an equation for the power available in wind:

$$\text{Power} = \tfrac{1}{2}\rho A v^3$$

The amount of power captured by a wind turbine is considerably less than the amount of power available in wind, since air must keep moving past the turbine. Wind cannot give up all of its kinetic energy to a turbine or the air would come to a standstill behind the turbine and pile up, blocking subsequent air. In theory, the maximum efficiency of an ideal wind turbine is just over 59%, first calculated by Albert Betz in 1919. The four sails of traditional European windmills achieved an efficiency of around 7%, while American windpumps operated at less than 4%. By contrast, the best turbines today capture more than 50% of the wind's power. A variety of factors affects the efficiency of wind turbines, including the shape and pitch of the rotor blades, the gearing mechanisms, and the type of generator.

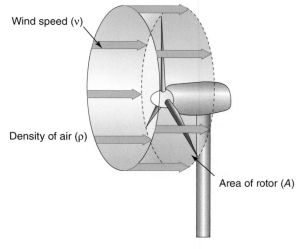

Wind speed (v)

Density of air (ρ)

Area of rotor (A)

FIGURE 8.8 A wind turbine extracts energy from a column of air whose base equals the area swept out by the rotor blades and whose length depends on the wind speed. The greater the wind speed, the longer the air column that strikes the blades each second. Denser air increases the impact force on the blades. Of the three factors—wind speed, rotor area, and density of air—wind speed is the most important in determining power available to a turbine.
© 2016 Cengage Learning®.

EXAMPLE 8.1

Suppose a wind turbine produces 10 kW of power during the summer when the wind is blowing at 10 mph.

a. How much power will the turbine produce in a summer breeze of 20 mph?
b. In the winter, when the air is 15% denser than during the summer, how much power will the turbine produce in a 10-mph breeze?
c. Suppose a turbine of identical design has rotor blades that are twice as long. How much power will this turbine produce in a 10-mph summer breeze?

Solution:

a. Even though the details of the particular turbine are not provided, the cubic relationship of the speed enables a comparison, assuming the other two parameters (rotor area and air density) remain constant. Dividing the power equation for the unknown power P_2 at 20 mph by that for the known power P_1 at 10 mph provides such a comparison.

$$P_2 / P_1 = \tfrac{1}{2}\rho A v_2^3 / \tfrac{1}{2}\rho A v_1^3$$
$$P_2 / P_1 = v_2^3 / v_1^3$$
$$P_2 / P_1 = \left(\frac{v_2}{v_1}\right)^3$$
$$P_2 / P_1 = (20/10)^3$$
$$P_2 / P_1 = 2^3 = 8$$
$$P_2 = 8\,P_1$$
$$P_2 = 8 \times 10 \text{ kW}$$
$$P_2 = 80 \text{ kW}$$

Doubling the wind speed increases the power output by a factor of eight. Hence, small changes in wind speed substantially impact the power generated.

b. A similar process allows for a comparison resulting from a change in air density (this time assuming the wind speed and rotor area remain constant). Dividing the power equation for the unknown power P_2 of the winter by that of the known power P_1 of the summer provides a solution.

$$P_2 / P_1 = \tfrac{1}{2}\rho_2 A v^3 / \tfrac{1}{2}\rho_1 A v^3$$
$$P_2 / P_1 = \rho_2 / \rho_1$$
$$P_2 / P_1 = 1.15/1$$
$$P_2 / P_1 = 1.15$$
$$P_2 = 1.15\,P_1$$
$$P_2 = 1.15 \times 10 \text{ kW}$$
$$P_2 = 11.5 \text{ kW}$$

The greater air density of winter increases output only slightly over the summer value.

c. Increasing the length of the rotor blades will increase the area intercepted by the rotor. Again dividing the power equation for the unknown power P_2 of the larger rotor by the power equation for the known power P_1 of the smaller rotor yields the solution.

$$P_2 / P_1 = \tfrac{1}{2}\rho A_2 v^3 / \tfrac{1}{2}\rho A_1 v^3$$
$$P_2 / P_1 = A_2 / A_1$$
$$P_2 / P_1 = \pi r_2^2 / \pi r_1^2$$
$$P_2 / P_1 = \left(\frac{r_2}{r_1}\right)^2$$

$$P_2 / P_1 = 2^2$$
$$P_2 / P_1 = 4$$
$$P_2 = 4 P_1$$
$$P_2 = 4 \times 10 \text{ kW}$$
$$P_2 = 40 \text{ kW}$$

Doubling the length of the rotor blades increases the area of the rotor by four times, which in turn increases the power output by the same amount.

EXAMPLE 8.2

Suppose two identical wind turbines are situated on identical towers, one in Wichita, Kansas, and one in Shawnee, Oklahoma. The site in Wichita experiences steady 20-mph winds for 8 hours; the site in Shawnee experiences 4 hours of wind at 10 mph followed by 4 hours of wind at 30 mph. Although each site offers an average wind speed of 20 mph over the 8-hour period, the available wind energy is not the same. Which site has more? By how much?

Solution:

The available energy (kWh) represents the power (kW) multiplied by the time (hours). For Wichita, this requires a single calculation over an 8-hour period. For Shawnee, it requires two calculations: one at 10 mph for the first 4 hours and one at 30 mph for the second 4 hours. Since the density of air and the area of the rotors are not provided, the solution can only be comparative and can be found by dividing the results for each of the two sites.

Wichita

$$\text{Energy} = \text{Power} \times \text{Time}$$
$$E_w = \tfrac{1}{2}\rho A_1 v^3 \times \text{Time}$$
$$E_w = \tfrac{1}{2}\rho A (20 \text{ mph})^3 \times 8 \text{ hr}$$
$$E_w = \tfrac{1}{2}\rho A (8000 \times 8)$$
$$E_w = \tfrac{1}{2}\rho A (64{,}000)$$

Shawnee

$$\text{Energy} = \text{Power}_1 \times \text{Time}_1 + \text{Power}_2 \times \text{Time}_2$$
$$E_S = \tfrac{1}{2}\rho A (10 \text{ mph})^3 \times 4 \text{ hr} + \tfrac{1}{2}\rho A (30 \text{ mph})^3 \times 4 \text{ hr}$$
$$E_S = \tfrac{1}{2}\rho A (1000 \times 4) + \tfrac{1}{2}\rho A (27{,}000 \times 4)$$
$$E_S = \tfrac{1}{2}\rho A (4000) + \tfrac{1}{2}\rho A (108{,}000)$$
$$E_S = \tfrac{1}{2}\rho A (112{,}000)$$

Dividing these equations provides a comparison between the two sites.

$$E_S/E_w = \tfrac{1}{2}\rho A (112{,}000)/\tfrac{1}{2}\rho A (64{,}000)$$
$$E_S/E_w = 112{,}000/64{,}000$$
$$E_S/E_w = 1.75$$

Over the 8-hour period, the site at Shawnee provides 75% more energy than the site at Wichita, although both sites have the same average wind speed. The 4-hour period of 30-mph winds provides nearly all of the energy gathered at Shawnee. This demonstrates the dramatic impact of high wind speeds on the total energy output of wind turbines.

Wind Quality

All types of wind turbines require high-quality wind to generate electricity. Wind quality, a measure of the suitability of wind conditions for energy generation, depends on two parameters:

- **Strength of wind:** Moderate wind speeds in the range of 10–40 mph provide suitable power without overfatiguing components. Lesser winds offer minimal power and may be unable to overcome internal friction within the turbine; greater winds can exert damaging forces that compel the turbine to shut down.
- Uniform flow: Wind turbines can utilize only steady wind that is free of turbulence and sudden changes in direction. Rotating blades are not adequately nimble to adjust to rapidly changing wind, and the kinetic energy of turbulent air is wasted: instead of producing electricity, turbulent wind simply fatigues the rotor and other components and results in premature turbine failure.

High-quality wind, therefore, is strong and steady, without turbulence, and is generally found only in undisturbed flow. Raised obstacles, such as trees and buildings, reduce wind speeds and create regions of turbulence that extend downwind 20 or more times the obstacle's height. Hence, a row of 25-foot-tall trees can produce turbulence unsuitable for wind turbines more than 500 feet away, as shown in Figure 8.9.

Even sites free of raised obstacles experience significant decreases in wind speed near the ground. Friction with the ground inherently retards wind, and the rougher the ground, the greater the reduction in wind speed.

Modern Wind Turbines

Modern wind turbines range in size from very small models mounted on sailboats (Figure 8.10) and remote cabins that charge batteries to very large commercial generators that are capable of producing 10 million watts of power or more (Figure 8.11). The blades of these turbines vary proportionally, with the smallest less than a foot long and the largest more than 300 feet long (see Figure 8.12).

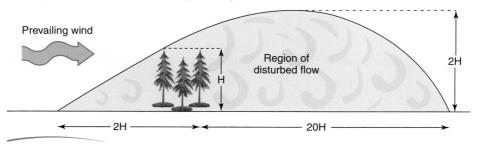

FIGURE 8.9 An obstacle, such as a row of trees, creates an envelope of disturbed air that extends downwind 20 or more times the height of the obstacle. This creates low-quality wind unsuitable for generating electricity. © 2016 Cengage Learning®.

FIGURE 8.10 A micro turbine mounted on a sailboat extracts some of the wind's energy for charging batteries. This turbine—using a 3-foot-diameter rotor—is rated to produce a peak output of 400 watts. © iStockphoto.com/John F Scott.

FIGURE 8.11 Commercial wind turbines, such as this one being installed in Montana, use rotors 225 feet in diameter and can generate more than a million watts of power. Each of these turbines supplies enough electricity to operate 500 residential homes. *Source:* Klaus Obel/NREL.

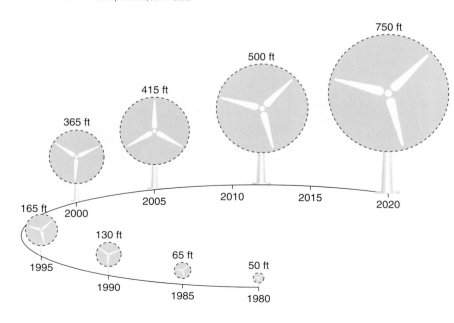

FIGURE 8.12 From micro turbines intended to charge batteries on sailboats and in cabins, to residential turbines that power a single home, to enormous commercial generators that produce enough electricity to run a small city, wind turbines vary radically in size.
© 2016 Cengage Learning®.

Despite the considerable range of sizes, all wind turbines—whether at the residential scale or the commercial scale—operate on the same principles. Nearly all modern wind turbines harness energy through the principle of lift rather than drag and fall into two basic categories: horizontal-axis wind turbines (where the blades rotate about a horizontal axis) and vertical-axis wind turbines (where the blades rotate about a vertical axis). While each type offers advantages, horizontal-axis wind turbines mounted atop tall towers make up the vast majority of all modern wind turbines, particularly at the commercial scale.

Horizontal-Axis Wind Turbines

Horizontal-axis wind turbines must face into the wind to produce power and, therefore, require some mechanism to constantly realign the rotor during changes in wind direction. For small turbines, this mechanism is typically just a tail vane (Figure 8.13a). For commercial-scale turbines, an electronically operated motor repositions the rotor according to wind sensors mounted on the housing (Figure 8.13b).

FIGURE 8.13 (a) The 11½-foot blades of this horizontal-axis wind turbine sweep out an area of over 400 square feet and can produce 10 kilowatts of power. This turbine provides power for the Tatoosh Island lighthouse off the coast of Washington. *Source:* Ed Kennell/NREL.

(b) This commercial-scale horizontal-axis wind turbine, part of the Maple Ridge Wind Farm, looms over a residential district in upstate New York. *Source:* PPM INC./NREL.

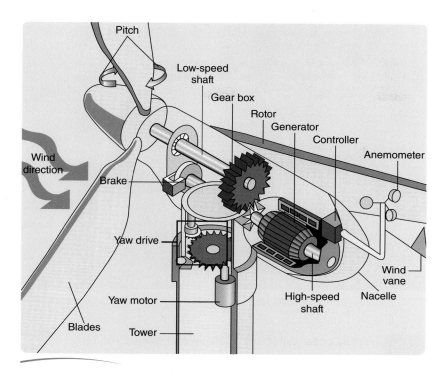

FIGURE 8.14 Most commercial wind turbines limit the rotation speed of the blades to reduce noise and improve performance. Gears convert the low-speed motion of the rotor shaft to the high-speed motion of the generator shaft. Most commercial wind turbines use variable-pitch blades to increase efficiency and rely on yaw drive motors to point the turbine into the wind. Electronic signals from a controller, which reads wind speed and direction from instruments mounted on the covering, regulate both the pitch and the yaw motors. © 2016 Cengage Learning®.

The blades of a horizontal-axis wind turbine move perpendicular to the direction of the wind, allowing them to harness energy throughout their entire rotation cycle. This makes them highly efficient at extracting wind energy.

Nearly all horizontal-axis turbines are mounted on top of tall towers, where the wind is stronger and less turbulent. The higher-quality wind allows the turbines to generate more power and minimizes turbine fatigue. However, locating all moving parts at the top of a tall tower makes installation and maintenance more difficult and expensive, and wind turbines require regular maintenance. This is particularly true of commercial-scale wind turbines due to their complexity (see Figure 8.14).

Vertical-Axis Wind Turbines

Vertical-axis wind turbines harness wind from any direction without repositioning the rotor. The absence of tail vanes or sensor-controlled motors simplifies their design. The gearbox and generator are located in the lower base of the turbine, making them much

FIGURE 8.15 These Darrieus-style vertical-axis turbines stand 60 feet tall at Altamont Pass in California and produce as much as 240 kW of power. © spirit of america/Shutterstock.com.

more accessible for maintenance. However, the design of these turbines limits options for tower mounting, particularly at the commercial scale. Consequently, most large vertical-axis wind turbines are mounted directly on the ground, where wind quality is compromised (Figure 8.15).

Vertical-axis wind turbines tend to be less efficient than horizontal-axis wind turbines because the blades backtrack against the wind for half of every cycle. Furthermore, the rotation toward and against the wind stream fatigues blades, which may compromise their longevity. In addition, due to symmetrical forces on the blades, many vertical-axis turbines will not begin rotating under operational wind speeds without first being started by an electric motor.

For these reasons, very few commercial-scale vertical-axis wind turbines are currently being manufactured. Residential-scale models, on the other hand, continue to intrigue customers with modern designs and captivating aesthetics. Furthermore, small-scale vertical-axis wind turbines outperform horizontal-axis turbines when mounted to rooftops amid the turbulent air that often surrounds buildings. As a result, vertical-axis wind turbines have found a niche in some urban areas (Figure 8.16).

See Table 8.1 and Table 8.2 for comparisons of various turbines.

FIGURE 8.16 Small-scale vertical-axis wind turbines, such as this 10-foot-diameter helical model rated at 6 kW, are better able to harness low-quality wind common to urban areas, making them popular near homes and commercial buildings. © iStockphoto.com/Kim Dailey.

TABLE 8.1 Comparison of horizontal-axis and vertical-axis wind turbines.

TYPE OF TURBINE	ADVANTAGES	DISADVANTAGES
Horizontal axis	• Blades harness energy throughout the entire rotation cycle. • Turbines are commonly mounted on tall towers, where the wind resource is greater. • Adjustable blade pitch offers superior control over a wide range of wind speeds. • Turbines are able to start rotating at low wind speeds.	• Turbines must be constantly repositioned to face into the wind. • The tower-mounted gearbox and generator reduce accessibility for maintenance. • The tower increases installation costs.
Vertical axis	• Turbines are able to harness wind from any direction without repositioning the rotor. • The ground-mounted gearbox and generator improve accessibility for maintenance. • Turbines are better able to harness turbulent air. • The absence of a tower reduces installation costs. • Turbines require smaller foundations and create less stress on the support structure, improving roof-mounting options.	• Blades backtrack against the wind, reducing efficiency. • Fixed blade pitch impairs the ability to control excessive speed during strong wind. • Sometimes turbines are not self-starting. • Blades are prone to fatigue, as they rotate toward and against the wind stream.

© 2016 Cengage Learning®.

TABLE 8.2 Comparison of common horizontal-axis wind turbines.

FEATURE	MICRO	RESIDENTIAL	MIDSIZED	COMMERCIAL
Rotor diameter	2–10 ft	10–25 ft	30–100 ft	200–600 ft
Rated power	Up to 1 kW	1–10 kW	20–250 kW	1.5–12 MW
Rated wind speed	20–30 mph	20–35 mph	20–30 mph	20–40 mph
Minimum tower height	30–100 ft	60–100 ft	100–200 ft	200–500 ft
Common speed control mechanism	Furling	Furling	Pitch/stall control	Pitch control
Repositioning mechanism	Tail vane	Tail vane	Yaw motor	Yaw motor
Installation cost (including tower)	$2000–$12,000	$18,000–$75,000	$125,000–$400,000	$2 million–$15 million

© 2016 Cengage Learning®.

Operational Limits

Modern wind turbines are designed to operate over a range of wind speeds. Each turbine has slightly different specifications, but all turbines must incorporate methods of limiting operation to safe speeds.

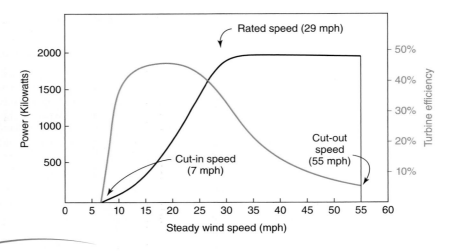

FIGURE 8.17 The power produced by a wind turbine varies from zero at its cut-in speed to a maximum at its rated speed. At wind speeds above the rated speed, the power output remains constant as the turbine sheds excess energy by becoming increasingly less efficient. This turbine has a rated power output of 2 million watts at a rated wind speed of 29 mph. © 2016 Cengage Learning®.

Friction and inertia within a turbine require a minimum wind speed to initiate rotation and produce power, called the **cut-in speed.** Winds less than the cut-in speed fail to rotate the turbine and do not generate electricity. Conversely, in very strong winds, a turbine may be driven beyond its structural limits, overloading the rotor, drive train, and generator. The maximum operational wind speed is called the **cut-out speed,** and the turbine must be designed to limit or cease operation above this level.

The **operational wind speeds** (the range between the cut-in and cut-out speeds) typically extend from 6 mph to 40 mph in residential-scale turbines and from 8 mph to 55 mph in commercial-scale turbines. Most commercial-scale turbines produce maximum power between 25 mph and 35 mph, called the **rated wind speed**, and shed excess energy at higher wind speeds (see Figure 8.17).

The rated power output of wind turbines occurs only when wind is blowing at or above the rated wind speed. In practice, this occurs only a fraction of the time. At the commercial scale, wind turbines typically operate at their rated power only 20%–40% of the time.

Depending on the size of the turbine, different methods are used to reduce excessive rotation rates during strong winds. Small wind turbines used in residential applications limit excess speed by turning out of the wind stream, either by tilting up or by rotating to the side. This process, called **furling**, is shown in Figure 8.18a. Turbines that furl are mounted slightly to one side of the tower so that, during strong winds, the pressure on the rotor forces the turbine to pivot (Figure 8.18b). Spring-loaded hinges, weights, or hydraulics determine the furling wind speed.

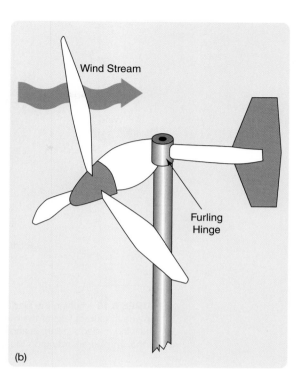

FIGURES 8.18 When a wind turbine furls, it pivots out of the wind stream. This greatly reduces its efficiency at extracting energy and prevents excessive rotation rates. **(a)** *Source: Doug Nelson/NREL.* **(b)** © 2016 Cengage Learning®.

Large wind turbines limit excess speed in any of three ways:

- **Pitch control:** Variable-pitch turbines alter the alignment of the blades with respect to the wind, turning them more parallel to the wind so that wind slips past without generating lift (Figure 8.19a). This method allows the greatest control in regulating the turbine's power output. Nearly all large modern turbines use pitch control.
- **Stall control:** Alternatively, variable-pitch turbines may alter the alignment of the blades so that they become more perpendicular to the wind direction. This process, called **stalling**, destroys lift by causing turbulence behind the blades (Figure 8.19b). This blade orientation maximizes the wind's force against the face of the blades, however, increasing the stress on the rotor and the wind load against the tower.
- **Passive stall:** For turbines with blades mounted at a fixed pitch, the blades can be designed to create turbulence naturally at excessive wind speeds, inducing a stalled situation. For this design, the pitch usually varies along the length of the blades, twisting from hub to tip, which induces the blades to stall gradually and prevents excessive wear.

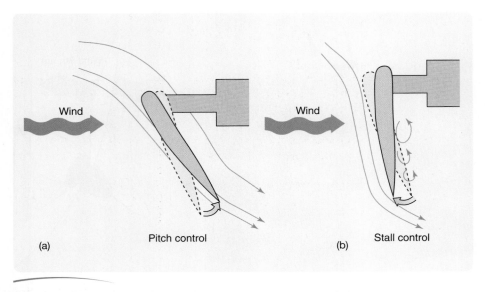

(a)   Pitch control   (b)   Stall control

FIGURES 8.19 Pitching the blade to become more parallel to the wind reduces lift and causes the blade to extract less of the wind's energy. Turning the blade to become more perpendicular to the wind, called stalling, is another method to limit the extracted energy. However, stalling can also maximize drag acting on the blade. © 2016 Cengage Learning®.

Siting

Selecting a suitable site for installing wind turbines requires information on prevailing wind speeds and patterns, locations of obstacles, proximity to habitation and electrical services, and other environmental issues.

Wind resource maps, such as the one shown in Figure 8.20, provide an estimate of the average annual wind speeds at various locations. These maps rarely indicate the distribution of wind speeds, however, which (as demonstrated in Example 8.2) is essential for accurately determining the site's actual wind energy potential. Local variations not indicated by resource maps can also be substantial. A formal site assessment usually entails monitoring wind speeds for several years and examining indicators provided by vegetation and other weathering patterns. The **Griggs-Putnam Index** (Figure 8.21), which links long-term prevailing wind with coniferous tree deformation, is an indicator sometimes used by site assessors.

Methods of estimating surface roughness of various terrains allow site assessors to extrapolate wind speeds for different heights when provided with a known value at a measured height. This helps determine appropriate tower height and overall suitability of the site.

Tall towers require considerable investment of materials and deep or massive foundations to anchor them solidly. The expense of erecting a tall tower generally makes residential wind power more expensive per unit of energy produced than commercial wind power. Nevertheless, wind turbines of all sizes offer a valuable source of clean energy and can be particularly valuable for off-grid homes, farms, and cabins and for remote applications.

FIGURE 8.20 The average annual wind speed measured 80 meters (260 feet) above the surface provides an estimate of the wind resource for the United States. © 2016 Cengage Learning®. *Source:* Adapted from AWS Truepower™ and the National Renewable Energy Laboratory (NREL).

Further Learning OCEAN CURRENT POWER

Shallow ocean currents are caused largely by winds blowing across the ocean surface and are influenced by the rotation and topography of the earth. Since the sun's uneven heating of the earth is the source of wind, ocean currents are another indirect form of solar energy.

Underwater turbines capture energy from ocean currents much like wind turbines capture energy from wind. Both convert the kinetic energy of a moving fluid—whether water or air—into electricity.

Power available in ocean currents, like that available in wind, depends on the speed of the current, the area swept out by the rotor, and the density of water, and it follows the same relationship: $P = \frac{1}{2}\rho A v^3$. Although ocean currents move slowly compared with typical wind speeds, the density of water is 830 times greater than that of air, so even moderate ocean currents contain enormous amounts of energy.

While people have considered harnessing ocean currents for power generation since the mid-1970s, no commercial grid-connected turbines currently operate. Countries such as the United States, Japan, China, United Kingdom, and Canada are presently testing prototypes. The ocean is a hostile environment for mechanical devices and poses several challenges. Foremost, designs must address and prevent the following problems:

- Corrosion can damage the support structures, anchoring cables, rotors, generators, and electrical cables.

- The buildup of marine growth, such as barnacles, muscles, and algae, can foul rotor blades and add resistance to the structures.

- The sudden formation and collapse of bubbles from the rotor blades, a process called cavitation,

(continued)

(continued)

results in damaging turbulence. Cavitation reduces the rotor's efficiency and contributes to fatigue and premature failure, just as turbulence does in wind turbines.

Ocean current turbines must also be designed with minimal maintenance requirements, prevent harm to fish and marine mammals, and offer no interference with shipping routes, fishing grounds, and recreational uses.

Despite difficulties imposed by a marine environment, the extraction of energy from ocean currents has notable advantages over that from wind:

- Water currents provide a much greater energy density than wind, allowing comparatively small turbines to generate large amounts of power.
- Ocean currents are much more steady than wind, providing predictable power that can serve as the base load for utility power.
- The submersion of turbines, while complicating maintenance, minimizes their visual impact—the dominant complaint about modern wind power.

A promising site for the deployment of underwater turbines in the United States includes the Florida Straits Current and Gulf Stream off the east coast of Florida. The U.S. Department of Energy estimates that each square foot of rotor area could extract about 100 watts of power and that capturing as little as 1/300 of the total available energy could supply Florida with all of its electricity needs.[1] However, some scientists are concerned that extracting even this amount of kinetic energy may lead to undesirable environmental consequences such as slowing the current, altering the path of the Gulf Stream, or changing flow patterns around nearby estuaries or other sensitive ecosystems. In any case, underwater turbines provide a promising complement to wind energy and may offer a low-impact source of renewable energy.

Environmental Impact

The environmental impact of wind power is remarkably small, particularly when compared with conventional methods of energy production using fossil fuels. The operation of wind turbines generates no carbon dioxide and releases no pollutants that can contribute to acid rain or smog. Wind power does not create radioactivity, release toxic compounds or heavy metals, or contaminate air or water resources.

Commercial wind power is currently the least expensive form of renewable energy and can be implemented more quickly and with less environmental impact than most other types of power-creating facilities. The energy generated by commercial turbines usually offsets the embodied energy of their construction and installation within a year or so of operation. Furthermore, wind towers can be placed on open fields used for farming or ranching with minimal impact and can provide income to landowners through lease and royalty agreements (Figure 8.22).

Many of the problems once associated with wind turbines, such as excessive noise, have been eliminated by slowing the rotation rate, redesigning the blades, and siting turbines away from residential areas. The sound levels observed a quarter of a mile away from modern commercial wind turbines are generally no greater than the sound levels within a library. Unfortunately, some small residential wind turbines that operate at high rotation rates continue to produce excessive noise, disturbing neighbors and resulting in negative opinions of wind power.

Index	Top View of Tree	Side View of Tree	Description of Tree	Wind Speed (mph)
0			No Deformity	No Significant Wind
I			Brushing and Slight Flagging	7–9
II			Slight Flagging	9–11
III			Moderate Flagging	11–13
IV			Complete Flagging	13–16
V			Partial Throwing	15–18
VI			Complete Throwing	16–21
VII			Carpeting	22+

FIGURE 8.21 Evergreen trees can serve as indicators of prevailing wind conditions, and studies of tree deformation with known prevailing winds created this table. Site assessors can use this information to gauge the wind resource of a site.
© 2016 Cengage Learning®.

FIGURE 8.22 Many farmers find it profitable to lease their land for commercial wind installations. The turbines minimally disrupt farming and grazing and can provide significant revenue. *Source:* Ruth Baranowski/NREL.

Modern wind turbines no longer pose exceptional hazards for birds. Despite lingering public concern, numerous studies of bird mortality clearly indicate that tall structures (buildings, towers, stacks, utility poles), as well as automobiles, transmission lines, and house cats, cause significantly more harm to birds than wind turbines. The slower rotation rates of turbine blades, along with the use of tubular towers that deny roosting opportunities, have greatly reduced bird mortality over the early designs that prompted the reputation.

Commercial wind turbines sited near airports or military zones may interfere with near-ground radar. They may also cause interference with radio and television signals if positioned between the transmitter and receiver. The rotating blades of wind turbines sometimes cast flickering shadows when backlit by the sun. Although potentially bothersome, all of these issues can be easily resolved by appropriate siting. Zoning regulations specific to commercial wind turbine installations are already common in many municipalities to circumvent these issues. Residential-scale wind turbines are rarely large enough to raise any of these issues.

The dominant environmental impact of wind turbines, particularly of commercial wind farms, is visual pollution (Figure 8.23). In rural settings, wind turbines have been accused of adding clutter and discord to the landscape. Unlike conventional power plants, which are usually urban, wind turbines may infringe on land that is otherwise considered scenic, creating a conflict of expectations. Situating turbines in an aesthetic arrangement and maintaining uniformity of

FIGURE 8.23 Critics of large wind farms claim the turbines are visually tiring and aesthetically disruptive. *Source:* PPM INC./NREL.

size, color, form, and rotation direction can mitigate some of this. For many, the sight of a few large wind turbines tends to disrupt the senses less than an abundance of small machines. Wind farms usually bury interconnecting power lines to minimize their industrial character and select paints that blend the towers and turbines with the skyline.

The installation of wind turbines offshore is a potential means of mitigating visual pollution, provided the farms are sufficiently distant that they cannot be seen from coastal communities (Figure 8.24). Offshore installations pose numerous challenges, however, including corrosive salt spray, ocean storms, extensive foundations, and power transmission from the turbines to onshore distribution centers. Nevertheless, not only do offshore installations remove turbines from public sight, but also the uniform surface of the ocean provides higher-quality winds, promoting the construction of larger, more powerful turbines.

FIGURE 8.24 Offshore wind farms may mitigate visual pollution by removing turbines from view. *Source:* Robert Thresher/NREL.

CASE STUDY 8.1

University of Minnesota–Morris

Location: Central Minnesota (45.5° N latitude)

Background: The University of Minnesota–Morris (UM–Morris), located on the prairies of west-central Minnesota, enrolls 1900 undergraduate students in a liberal arts curriculum that collaborates with the University of Minnesota's West Central Research and Outreach Center (WCROC). In 2005, the WCROC installed a 1.65 MW Vestas V82 wind turbine to generate electricity for the center and the UM–Morris campus. In February 2011, a second identical turbine—this time funded by UM–Morris itself—began operations (Figure 8.25). The two wind turbines support the university's commitment to generate all of its own energy and become carbon-neutral. Toward this goal, the campus also installed a biomass gasification facility that converts corncobs and prairie grass residue into heat and electricity. Furthermore, a ground-mounted system of 32 flat-plate solar hot water collectors heats its recreational swimming pool, and a 2.6-kW photovoltaic array provides additional electricity.

Description: The Vestas V82 is a three-bladed horizontal-axis wind turbine with a rotor diameter of 270 feet. The turbine sits atop a 230-foot-tall tubular tower and operates between a cut-in speed of 8 mph and a cut-out speed of 45 mph. At its rated speed of

FIGURE 8.25 Courtesy of Kari Adams, Graphic Designer, University of Minnesota, Morris.

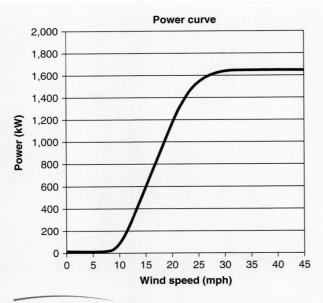

Power curve

FIGURE 8.26 © 2016 Cengage Learning®.

29 mph, the turbine generates 1650 kW of electricity (see Figure 8.26). Independent pitch control on each of the three blades limits excessive rotation rates during high wind speeds. The turbine maintains a rotation rate of roughly 14 revolutions per minute throughout its operation. The UM–Morris turbine experiences an average wind speed of 16 mph.

Cost Analysis: The 2011 installation of the wind turbine owned by UM–Morris cost $4.4 million. The university also pays an additional $67,000 per year to service and maintain the turbine. A contract with the local utility company promises revenue of $0.08 per kWh of electricity delivered to the grid. Because this wind turbine produces an average of 450,000 kWh of energy per month, it generates revenue of $36,000 per month. This amounts to over $430,000 per year and represents an 8% return on investment, paying for itself in just over 12 years. Figure 8.27 shows the electrical energy output of the two wind turbines operating at UM–Morris. The largest total energy output (beginning in March 2011) indicates the addition of the wind turbine owned by UM–Morris.

Although the electricity generated by the two wind turbines exceeds the campus electricity demands by about 1 million kWh each year, the campus must still buy electricity from the utility company because the bulk of the electricity generated by the turbines occurs at night, when campus demand is low. During the night, excess wind-generated electricity is sold to the utility at off-peak prices, only to be repurchased during the daytime at the higher peak price. On average, the wind turbines provide all of the electricity used on campus only 47% of the time. For this reason, future installations of renewable energy systems will focus on daytime production—for example, expanding the existing photovoltaic array or increasing electricity generation from biomass.

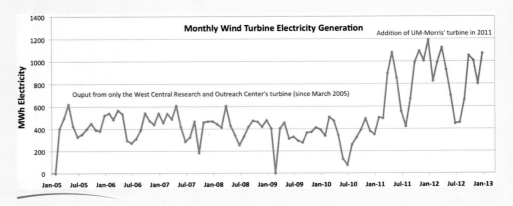

FIGURE 8.27 Courtesy of Kari Adams, Graphic Designer, University of Minnesota, Morris.

Chapter Summary

Human society has harnessed the power of wind since antiquity, and fashioning machines to generate electricity follows a long history of grinding grain and pumping water. Although drag-based designs have captivated the interest of inventors through the ages, all modern wind turbines apply the principles of aerodynamic lift to extract energy from wind. As a result of advancements in technology and materials science, modern wind turbines use strong, lightweight, precision-crafted blades; low-friction drive trains; sophisticated blade-pitching mechanisms; and highly efficient generators to reduce maintenance and maximize output.

To generate electricity, wind turbines require steady, nonturbulent air, and the output power depends heavily on wind speed, making small changes in the wind resource highly significant. Compared to other forms of power generation, the environmental impact of wind turbines is remarkably small—less than most other forms of renewable energy and far less than conventional power generation. Wind energy's greatest impact is visual, and raising awareness of its many benefits to society—such as pollution-free energy, opportunities for economic development, and revenue to communities, farms, and individuals—can lessen this drawback.

Review Questions

1. Commercial wind power is presently the fastest-growing source of renewable energy. Why do you think wind power is preferred over solar power at the commercial scale?
2. Consider residential wind and solar power, comparing each of the following:
   a. Site requirements
   b. Initial expense
   c. Long-term cost of energy production
   d. Maintenance requirements
   e. Ability to add to system after original installation
   f. Compatibility with off-grid and grid-tied applications
   g. Building integration options
3. What are the primary differences between residential- and commercial-scale wind turbines? How does this contrast with residential- and commercial-scale photovoltaics?
4. In what ways do wind and solar electricity complement each other? Why is this particularly effective for off-grid applications?
5. A wind turbine installed in a windy pass high in the Andes Mountains is observed to produce less than its rated power, despite being appropriately sized for the wind resource. Give two reasons why this might be the case.
6. Windmills that ground grain during the sixteenth century were much more common in regions that lacked rapidly moving streams and rivers. Why was this the case?
7. Do the four long sails of the traditional Dutch windmill represent drag-based or lift-based design? What type of design is utilized by the many vanes of the American windpump?
8. An anemometer is a drag-based device. Why has it not been replaced with a more efficient lift-based alternative?
9. Under which condition would a wind turbine generate the greatest amount of power: a steady 10-mph wind or a highly turbulent 20-mph wind? Why?
10. Household wind chimes and rotating yard decorations are typically drag-based devices. For these applications, why is a drag-based design preferred over a lift-based design? Consider more than one explanation.
11. Why do many residential wind turbines but no large commercial wind turbines rely on furling to limit operation during excessive wind speeds?
12. Use the solar resource map (Figure 7.15) and the wind resource map (Figure 8.20) to determine the best locations within the United States to harness both solar and wind power.

Practice Problems

1. A wind turbine in Kansas has blades half as long as a wind turbine in Nebraska but experiences twice the wind speed. If all other conditions are equal, which turbine produces more power? By how much?

2. A wind turbine with a cut-in speed of 6 mph and a cut-out speed of 36 mph generates 1.6 kW at its rated speed of 20 mph. Determine the total energy produced over a 24-hour period given the following wind speeds:
   a. 5 mph for 6 hours
   b. 10 mph for 6 hours
   c. 20 mph for 6 hours
   d. 40 mph for 6 hours

3. A wind turbine in Montana experiences 25 mph winds in the summer and 23 mph winds in the winter when the air density is 15% greater. During what time of year does the turbine produce more power, summer or winter? By how much?

4. A commercial turbine with a blade length of 220 feet and a residential turbine with a blade length of 10 feet stand atop towers on the same hill and experience the same wind conditions. If the residential turbine produces 10 kW of power, how much will the commercial turbine produce? Assume both operate at the same efficiency.

5. A wind turbine located outside of Dodge City, Kansas, produces a maximum output of 3 kW at a rated wind speed of 25 mph.
   a. Estimate the average wind speed at this location from Figure 8.20.
   b. Calculate the power output using the average wind speed.

   c. Determine the energy generated (kWh) over the course of a year if the average wind speed occurs 30% of the time.

6. Determine the energy generated (kWh) over the course of a year for the situation described in Problem 5 if the actual wind speed consists of half the average wind speed for half the time and 50% more than the average wind speed for the other half. Again assume the wind blows for a total of 30% of the time.

7. A commercial wind turbine intended for off-shore installation sits atop a 400-foot-tall tower and generates 10 MW of power at a rated wind speed of 25 mph. It has a cut-in speed of 9 mph and a cut-out speed of 65 mph. The turbine controls excessive wind by pitching its three blades, each of which is longer than a football field, and maintains maximum power between its rated speed and its cut-out speed.
   a. Determine the energy (kWh) produced during a year by an offshore wind farm consisting of 20 of these turbines if they operate at their rated power for 40% of the time.
   b. If electricity sells for $0.12 per kWh, what revenue does this wind farm generate each year?
   c. If the installation of each turbine costs $12 million, how long will it take to pay off the initial investment?
   d. How much power does one of these turbines produce at its cut-in speed?
   e. What is the efficiency of the turbine at extracting power from the wind during a 60 mph gale?

Endnote

1 U.S. Department of the Interior, Minerals Management Service, Renewable Energy and Alternate Use Program. (2006). *Technology white paper on ocean current energy potential on the U.S. Outer Continental Shelf*. Washington, DC.

Hydropower

© Crady von Pawlak/Flickr/Getty Images.

Introduction

Hydropower was one of the first forms of mechanical assistance adopted by early civilizations. As early as 200 B.C., people applied waterwheels to crush grain and pump water. Waterwheels, like windmills, increased productivity and reduced dependence on human and animal muscle power. Wherever they were installed, waterwheels became engines for driving early technological advancement. The ancient Greeks used waterwheels to grind wheat into flour, and by the end of the Roman era, waterwheels provided a means to tan leather, smelt and shape iron, saw wood, prepare textiles, make paper, and impart mechanical assistance for many other industrial-type processes.

Since waterpower established a means for greater productivity and diminished manual labor, locations that offered good waterpower resources naturally became centers of industrial and economic activity. All over the world, many modern towns and cities owe their founding to the presence of waterpower.

During the eighth century, England had more than 5000 waterwheels in operation, roughly one mill for every 400 people. By the eighteenth century, waterwheels throughout Europe provided extensive power for all forms of manufacturing, including textile plants, concrete factories, and lumber mills. In addition, waterwheels operated heavy machinery such as cableways, draglines, and giant shovels. Waterwheels facilitated the colonization of America, providing many of the same forms of mechanical assistance employed in Europe (Figure 9.1).

FIGURE 9.1 Waterwheels, like this mill built in Virginia in 1905, once dotted the countryside and provided a vital form of mechanical power. © Mary Terriberry/www.Shutterstock.com.

Hydropower, unlike solar and wind, offers predictable and reliable energy. Despite the transformational impact of the coal-powered steam engine, which drove the Industrial Revolution in the late 1800s, hydropower remained important. Burning coal requires time to generate steam and cannot be turned off quickly, making it slow to respond to changing demands. Hydropower can be started and stopped almost instantly by regulating the flow of water. This flexibility carried it forward as the world shifted to fossil fuels, and as technology improved, hydropower incorporated the advancements and improved with it.

In 1849, James Francis developed a radial turbine that extracted energy from water with an efficiency of more than 90%. In 1882, Appleton, Wisconsin, became the site of the first hydroelectric power plant, which generated as much as 12.5 kW of direct current (DC) electricity and powered nearby paper mills (Figure 9.2). In 1890, a hydroelectric power plant in Oregon City, Oregon, successfully transmitted the first long-distance alternating current (AC) electricity. By the early 1900s, hydropower accounted for more than 40% of the United States' electricity production, and as late as the 1940s, hydropower supplied 75% of all the electricity consumed in the western United States.

FIGURE 9.2 A dam built across the Fox River in Appleton, Wisconsin, in 1882 supplied water for the world's first hydroelectric power plant. *Source:* Library of Congress, Prints & Photographs Division, LC-D4-4783 DLC.

World War II increased the United States' demand for electricity, and hydroelectric power plants provided a means to meet that need quickly. By 1944, large dams throughout the West had quadrupled the hydroelectric power output, supplying reliable energy to operate steel mills, oil refineries, automotive and aircraft factories, shipyards, and farm irrigation systems and to serve other purposes.

The amount of electricity generated from hydropower continued to increase until the 1970s, when environmental legislation imposed regulations that curtailed construction of large dams. While the electrical output of hydropower has remained relatively constant since then, the capacity of coal-fired power plants has steadily increased, reducing the relative contribution of hydropower from more than 40% in the late 1940s to only 7% today.

Further Learning   EARLY WATERWHEELS

Two principal types of waterwheels arose early in the history of waterpower and remained in use at the end of the eighteenth century. These were the undershot wheel and the overshot wheel.

- **Undershot wheel**: Numerous blades dip into a flowing stream, and the water pressure against the blades rotates the wheel. Although relatively inefficient, this type of waterwheel has the advantage of being able to operate in nearly any stream or channel, even those with relatively gentle gradients.

- **Overshot wheel**: Water falling from above strikes buckets and rotates the wheel using both the impulse of the water and its weight. This type of waterwheel is more efficient than the undershot wheel but could be applied only in locations that offered a vertical drop (head) of at least the diameter of the wheel.

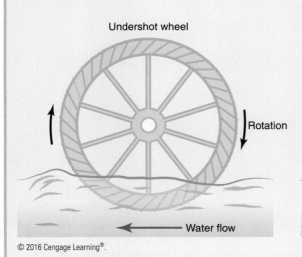

Undershot wheel

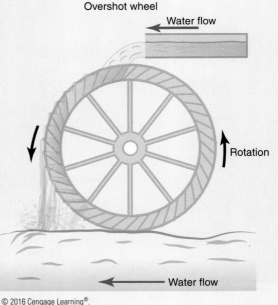

Overshot wheel

Power from Water

Hydropower converts the gravitational potential energy of water into electricity by rotating a turbine connected to an electric generator. Whereas wind can give up only a portion of its kinetic energy and offers no potential energy, water can completely forfeit both forms of energy. A hydroelectric turbine can fully arrest the flow of water, and afterward gravity draws the water away, allowing continuous extraction of power and making hydropower the most efficient of all commercial energy conversion methods.

The power available from water depends on two factors: the flow rate and the head.

- Flow rate (Q) determines the volume of water that arrives at the turbine each second, measured in gallons per second (gal/s).
- Head (H) measures the height in feet (ft) from which water descends and determines its pressure: the greater the head, the higher the water pressure arriving at the turbine.

These two variables—along with the density of water, ρ, and the acceleration of gravity, g—create an equation for the power, P, available in water:

$$P = \rho g Q H$$

$$P = 11.3 \frac{\text{watts}}{(\text{gal/s})\text{ft}} QH$$

Friction within the system and imperfections in turbine alignment and rotation speed reduce the actual power output to about 90% of this value. Small-scale hydropower installations that rely on long pipes to convey water to the turbine experience additional losses due to friction within the pipes and may achieve efficiencies of only 60% or less. By comparison, conventional fossil fuel and nuclear power plants perform at 35%–40% efficiency or less.

EXAMPLE 9.1

A pipe conducts 30 gallons of water per second from a diversion dam on a mountain stream to a hydroelectric turbine located in a valley 350 feet below. Assume friction within the pipe and turbine reduces the efficiency by 16% of its theoretical value. Calculate the turbine's power output P (in kW).

Solution:

The actual power produced is only 84% of the theoretical value.

$$P = 84\% \times \left(11.3 \frac{\text{watts}}{(\text{gal/s})\text{ft}} \right) \times Q \times H$$

$$P = 0.84 \times \left(11.3 \frac{\text{watts}}{(\text{gal/s})\text{ft}} \right) \times (30 \text{ gal/s}) \times (350 \text{ ft})$$

$$P = 100 \text{ kW}$$

A hydroelectric power plant of this size could meet all the power needs of a small community.

Since the power produced by a hydroelectric facility depends on both the flow rate and the head, the same amount of power can be generated under different circumstances (Figure 9.3). The design of the facility, as well as the type of turbine, depends on the specifics of the site. A site offering low head but high volume will appear quite different from one offering high head but low volume. Facilities are usually classified as low-head, medium-head, or high-head hydroelectric power plants, as shown in Figure 9.4. Differing degrees of topographic relief produce very different hydroelectric facilities (Figure 9.5–Figure 9.7).

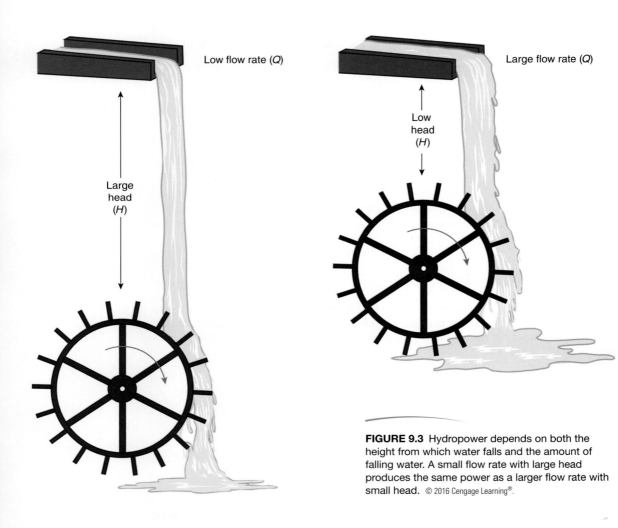

Low flow rate (Q)

Large head (H)

Large flow rate (Q)

Low head (H)

FIGURE 9.3 Hydropower depends on both the height from which water falls and the amount of falling water. A small flow rate with large head produces the same power as a larger flow rate with small head. © 2016 Cengage Learning®.

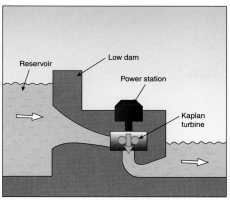

(a) Low head

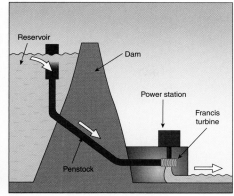

(b) Medium head

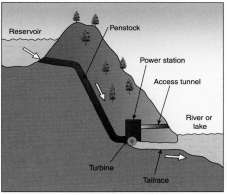

(c) High head

FIGURE 9.4 The type of hydroelectric facility depends on the conditions of the site and is classified as low, medium, or high head.

© 2016 Cengage Learning®.

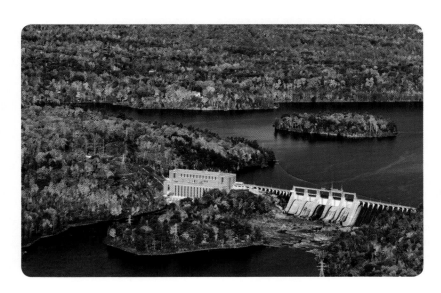

FIGURE 9.5 The Farmer Dam, located on the Gatineau River in Quebec, is a low-head facility that generates 100 MW of electrical power using a head of 65 feet.

© Pete Ryan/National Geographic/Getty Images.

FIGURE 9.6 The Hoover Dam, located on the Colorado River, is a medium-head hydropower facility that generates 2000 MW of electrical power with an average head of 520 feet. The energy production of the Hoover Dam averages 4.2 billion kWh per year. *Source:* Library of Congress, Prints & Photographs Division, photograph by Carol M. Highsmith, LC-USZ62-104919.

FIGURE 9.7 This hydroelectric station, located on Trollfjord in Norway, receives water from a mountain lake located almost 1500 feet above the generators. Norway produces the vast majority of its electricity from hydropower, most of which comes from high-head mountain lakes. Courtesy of Scott Grinnell.

Modern Water Turbines

Modern hydroelectric turbines come in a variety of designs, depending on the type of facility, the amount of head, and the flow rate. Turbines range in diameter from only a few inches (used to generate 100 watts) to more than 30 feet (for large commercial turbines producing many millions of watts). Two fundamental types of turbines exist: those that extract energy through reaction and those that extract energy through impact.

Reaction Turbines

Reaction turbines are completely submerged and entirely enclosed in a housing. Water arrives at the turbine under pressure and leaves the turbine at a much lower pressure. The pressure drop across the turbine accounts for much of the extracted power. The turbine uses curved blades that deflect the flow of water. In being deflected, water exerts a reaction force on the blades, causing them to rotate. Most commercial power plants use reaction turbines, designing the turbines to permit fish and other small objects to pass through unharmed. The two most common types of reaction turbines are the Francis and Kaplan turbines.

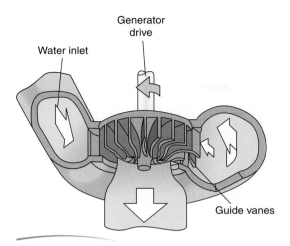

FIGURE 9.8 (a) A Francis turbine resides at the center of a curved tube. Water passes through precisely adjusted guide vanes, gives up its energy to the turbine, and flows away through the center draft tube.
© 2016 Cengage Learning®. *Source:* Adapted from an illustration by David Darling.

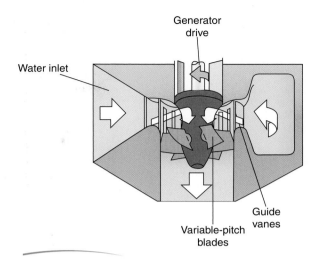

FIGURE 9.8 (b) Water flows down the axis of a Kaplan turbine, rotating its propeller-like blades much like air past a wind turbine, except a Kaplan turbine is mounted vertically to better extract gravitational potential energy.
© 2016 Cengage Learning®. *Source:* Adapted from an illustration by David Darling.

- **Francis turbines** (Figure 9.8a) are by far the most widely used type of turbine in commercial hydroelectric power plants. They perform efficiently in applications offering medium to high head and moderate flow rates. Water flows through a curved tube around the turbine, and guide vanes direct it radially inward. It then strikes the turbine blades, gives up most of its energy, and flows out through a central draft tube. Adjusting the guide vanes allows efficient operation over a range of flow rates.
- **Kaplan turbines** (Figure 9.8b) produce power efficiently in low-head, high-flow applications, where Francis turbines perform poorly. They resemble variable-pitch propellers mounted vertically so that the downward flow of water rotates the blades. Varying the pitch of the blades optimizes efficiency during changing flow rates.

Impact Turbines

Impact turbines are not submerged but rather use nozzles to focus high-pressure jets of water into curved bowls on the rim of a wheel. The impact of water striking the bowls causes the wheel to rotate. Under ideal circumstances, the wheel extracts all of the jet's kinetic energy, and the water falls away under the influence of gravity. Adjusting the size and number of jets regulates the power output. The most common type of impact turbine is called a **Pelton wheel** (Figure 9.9), named after the inventor who developed it in the 1870s. These turbines are used for low-flow facilities with high head and are common in residential and small-scale commercial systems.

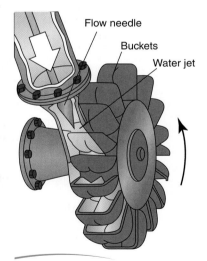

FIGURE 9.9 The bowl-shaped buckets of a Pelton wheel arrest the motion of the water from the jets; the water then falls away and flows out through the tailrace. The flow rate and alignment of the jets can be adjusted to maximize performance. © 2016 Cengage Learning®. *Source:* Adapted from an illustration by David Darling.

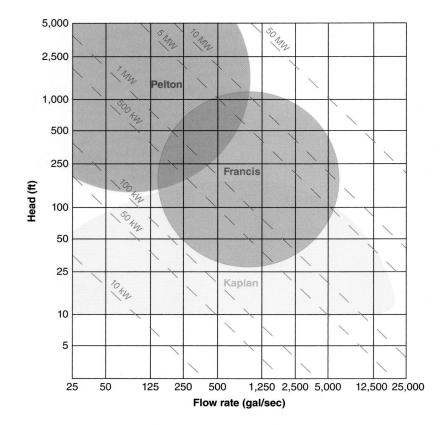

FIGURE 9.10 The Francis, Kaplan, and Pelton turbines perform optimally under different conditions. The Francis turbine is broadly applicable and used for most commercial hydropower installations. The Kaplan turbine outperforms the others for low-head, high-flow conditions, while the Pelton wheel serves best in situations of high head.

© 2016 Cengage Learning®.

Each of these three types of turbines—Francis, Kaplan, and Pelton—yields efficiencies greater than 90% when performing optimally. Selecting the appropriate turbine depends on the available head and flow rate, as shown in Figure 9.10.

Hydroelectric Systems

All hydroelectric systems consist of four basic components, as shown in Figure 9.11:

- **Water intake**: This varies with scale from an unobtrusive pipe set in a mountain stream to a large reservoir created by damming an entire river valley. Water flows through an intake screen that prevents large objects from entering the turbine (Figure 9.12).
- **Penstock**: The penstock is a channel or pipe that connects the water intake to the turbine and often includes a control gate for regulating flow rate (Figure 9.13).
- **Powerhouse**: The powerhouse encloses and protects the turbine, generator, and regulatory controls for power generation (Figure 9.14).
- **Tailrace**: The tailrace provides a means for water to leave quickly and easily without backing up or interfering with the turbine.

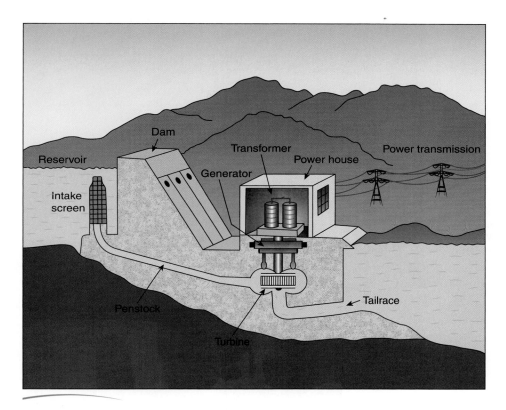

FIGURE 9.11 The components of a commercial hydropower plant. © 2016 Cengage Learning®.

FIGURE 9.12 The intake towers of the Hoover Dam screen large debris from entering the penstock and can accommodate considerable variations in reservoir level. *Source:* U.S. Department of the Interior, Bureau of Reclamation.

FIGURE 9.13 These six pipes convey water from the Upper Derwent Catchment in Tasmania and serve as the penstock for a 90-MW hydropower facility. Six Pelton wheels operate under a head of 950 feet. © Neale Cousland/www. Shutterstock.com.

FIGURE 9.14 A row of generators, each capable of producing over 120 MW of electrical power, is part of the Hoover Dam hydroelectric facility.

Source: U.S. Department of the Interior, Bureau of Reclamation.

Residential- and commercial-scale systems use similar turbine designs, selected according to the available head and flow rate, and operate under the same principles. The primary difference between the two is that commercial-scale systems generally make use of dams to create large reservoirs. The environmental impact of these dams can be considerable. Small-scale hydropower, called micro hydro, typically excludes dams and reservoirs, instead using barriers that divert a portion of the flow into a channel or pipe, as shown in Figure 9.15. Micro hydro results in minimal environmental impact.

Micro Hydro

Micro hydro systems range in capacity from less than 100 watts to more than 100 kW. Some provide power for a single home or off-grid cabin; others serve entire communities. Without the stored resources of a reservoir, the power produced by micro hydro installations varies with the seasonal flow of the water supply. In the United States, mountain streams often depend on springs and melting snow, which can result in reduced flow during the fall and winter. Nevertheless, this variability tends to be seasonal, rather than daily, which makes micro hydro more predictable than solar or wind. In many places around the world, micro hydro provides reliable power that operates day and night all year long without interruption. For this reason, micro hydro is often selected over solar or wind wherever suitable streams are available.

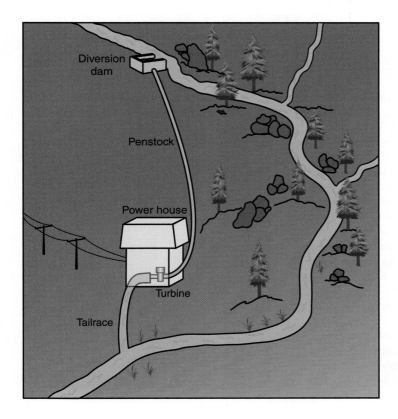

FIGURE 9.15 Micro hydro installations typically divert only a small portion of a river or stream and have minimal environmental impact. © 2016 Cengage Learning®.

For off-grid systems, micro hydro provides a stable and reliable source of energy. The constancy and predictability of hydropower, even if the available resource is small, dramatically improve off-grid systems that also incorporate more variable sources such as wind or solar. Since hydroelectric turbines produce power around the clock—whenever water is flowing—some load must exist to use that electricity, whether to charge a battery bank or to operate electrical equipment. Otherwise, damage may occur to the turbine and generator. Consequently, most off-grid hydroelectric installations include a diversion load, such as an electric resistance heater, that makes use of surplus electricity and minimizes the need to regulate water flow to the turbine.

Micro hydro generators operate well in an array of specialized applications as well, such as supplying electricity on sailboats. Any time the boat is in motion (or currents move past an anchored boat), hydroelectric generators mounted on the boat's hull produce electricity to operate lights, charge batteries, or run navigational equipment.

CASE STUDY 9.1

Holden Village Hydro Facility

Location: Northern Washington (48° N latitude)

Type of System	Flow Rate (gallons per second)	Head (feet)	Power Output (kW)	Original Date of Installation
Pelton wheel	11–53	640	66–227	1964

Description of System: Holden Village is a remote community located in the Cascade Mountains of Washington. Originally a small mining town, the village currently operates as a church retreat center, supporting as few as 50 inhabitants in the winter and as many as 400 or more in the summer. Holden Village includes six 50-person dormitories and 14 chalets, as well as a dining hall, kitchen, hospital, post office, school, and recreation center (with a bowling alley and full-sized basketball court). No roads connect the village with outside highways. Reaching the village requires a 36-mile boat journey from Chelan, Washington, and then an 11-mile bus ride up a winding dirt road that ascends 2100 feet. A micro hydro facility, consisting of two Pelton wheel turbines and generators, provides electricity for the community. The remote location prevents connection with the national utility grid, and mountain ranges severely restrict solar and wind resources, leaving hydropower as the most practicable form of energy.

Courtesy of Scott Grinnell.

Courtesy of Holden Village Operations.   Courtesy of Scott Grinnell.

The village and the hydroelectric facility are at an elevation of 3255 feet, located in a valley that is surrounded on all sides by towering mountains. A small dam at an elevation of 3895 feet diverts water from one of the many creeks. A 14-inch-diameter steel penstock delivers water to the hydro plant, dropping a vertical distance of 640 feet over a pipe length of roughly 2700 feet. Only a small portion of the creek is diverted by the dam, and the water returns to its natural drainage after emerging from the hydro plant. The environmental impact of the operation is extremely small.

The electrical output of Holden's hydro facility depends on the seasonal flow rate. The primary source of water for the mountain creeks is melting snow, which is the greatest during the summer and the least during early spring. A plot of the average power production for the years 1995–2011 reveals a maximum summer generation of 227 kW and a minimum spring generation of 66 kW. This seasonal variation prompts conservation during the spring but supplies ample power for the summer, when the village fills to its full capacity.

The hydroelectric facility generates standard three-phase AC power that is transmitted throughout the village at 2400 volts on conventional power poles. Local transformers reduce the voltage to 240 volts for each building.

Centralized boilers burning firewood harvested from the surrounding forest supplies heat for buildings through the winter. Otherwise, the hydroelectric plant satisfies all the energy needs of the community.

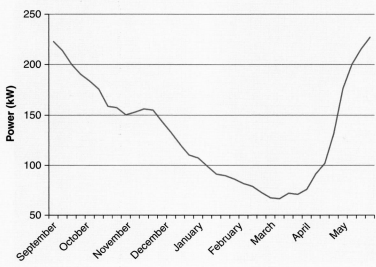

© 2016 Cengage Learning®.

CASE STUDY 9.2

Off-Grid Micro Hydro

Location: Northern Wisconsin (46.5° N latitude)

Type of System	Flow Rate (gallons per second)	Head (feet)	Power Output Average (watts)	Date of Installation
Low-head Turgo wheel	1.25	11	21	2003

Description of System: The outflow from a ¾-acre pond on an off-grid homestead powers a micro hydro system. Springs feeding the pond produce a steady flow rate of about 1.25 gallons per second, and the gently sloping terrain offers a head of 11 feet. The micro hydro system generates an average of 21 watts of continuous power and supplements a 950-watt photovoltaic array.

Pond water flows through a vertical inlet mounted 4 feet below the surface and continues through a sloping 6-inch-diameter pipe to a boxed enclosure a short distance downslope from the pond. Within the enclosure, the main pipe splits into two 3-inch-diameter pipes that feed water to two nozzles, one on either side of the turbine. The nozzles focus jets of water that rotate the turbine and generate electricity.

The turbine uses a small Turgo wheel, which is an impact wheel similar to a Pelton wheel but designed for small systems and able to accommodate low head. The turbine operates continuously day and night, producing power roughly 300 days per year and generating an annual output of about 150 kWh of electricity. While modest, this continuous supply of power provides stability to the uncertain energy received from the photovoltaic array and helps

Courtesy of Scott Grinnell.

Courtesy of Scott Grinnell.

maintain charge on the limited battery bank. When the batteries are full, excess electricity is automatically diverted through a resistance heater sheltered on the outside of the building.

Cost Analysis: The installation—including the turbine, controller, pipes, and other hardware; a 300-foot electrical line between the turbine and the residence; and a small heater to act as a diversion load when the batteries are at full capacity—cost $3700. At $0.12 per kWh, the electricity generated from the turbine saves about $18 per year. In this situation, the owners chose to invest in micro hydro for reasons of stability rather than economic returns.

Commercial Hydro

Commercial hydroelectric power plants range in capacity from less than 1000 kW to more than a million kW. Commercial plants tend to use hydropower differently than small-scale facilities: rather than supplying the base-load electricity, many commercial plants utilize hydropower primarily to provide supplementary power during periods of peak demand (Figure 9.16).

Hydroelectric generators can be started and stopped quickly, making them extraordinarily responsive to rapid changes in electrical demand. The gravitational potential energy of water stored in reservoirs can be converted into electricity at any time with very little advanced notice. Since fossil fuel and nuclear power plants generate electricity by producing high-pressure steam, they require relatively long start-up periods and are most efficient only at providing steady base loads. For this reason, hydropower plays an essential role in the nation's utility grid by providing quick responses to electrical demands that are difficult to achieve with conventional power plants.

A process called **pumped storage** increases the capacity of hydroelectric plants (Figure 9.17). When demand falls, such as during the night, surplus power from conventional

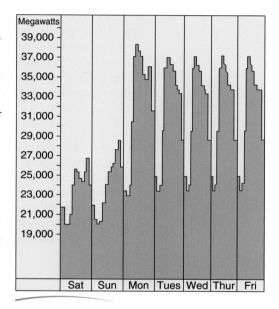

FIGURE 9.16 Electrical demand varies considerably over a typical week. Hydropower plants are much more efficient at accommodating these variations than steam power plants. © 2016 Cengage Learning®. *Source:* U.S. Department of the Interior, Bureau of Reclamation.

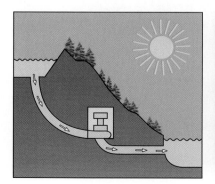 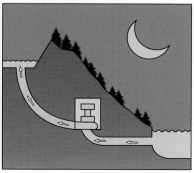

FIGURE 9.17 Pumped storage is a technique for maximizing hydropower's output during peak loads. Water that flowed through the turbines to produce power when needed during the day is pumped back into the reservoir at night so that it will be available again the next day. © 2016 Cengage Learning®. *Source:* U.S. Department of the Interior, Bureau of Reclamation.

steam power plants pumps water that has already flowed through the hydroelectric turbines back into the reservoir. The next day, when demand is high, the hydroelectric plant can reuse this water to create power. Despite the expenditure of energy to drive the pumps, this form of pumped storage is more efficient than relying on steam power plants to provide electricity for peak loads.

Environmental Impacts

The environmental impacts of commercial-scale hydropower are generally associated with the creation of dams (Figure 9.18). While there are hundreds of large hydropower dams in the United States, none has been built since 1979, when the 30-MW New Melones facility was completed in California. Environmental legislation passed in the late 1960s and 1970s made obtaining permits for large projects more difficult. However, by that time, nearly all major rivers in the United States had already been dammed, usually at multiple locations, and the most promising hydroelectric sites had already been utilized.

The environmental impacts of large-scale hydroelectric power plants are numerous.

- Dams can inundate productive land, destroy wildlife habitat, interrupt fish migration, change river ecology, compromise water quality, displace local populations, flood archaeological sites, ruin valuable agricultural land, and fragment river ecosystems.
- The release of water timed to accommodate peak demands in electricity bears no resemblance to natural flood cycles and can dramatically alter the sedimentation and erosion patterns downstream.

FIGURE 9.18 The environmental impact of dams can be considerable. Loggers removed these trees prior to flooding the valley, and drought re-exposed the stumps. © Nivek Neslo/Photodisc/Getty Images.

- The fertility of many river valleys around the world relies on periodic flooding to deposit silt and nutrients. Hydroelectric dams situated upstream of these valleys have largely eliminated this type of flooding, replacing the natural agricultural system used for millennia with one relying on fertilizers and irrigation.
- Water released from dams is usually warmer than the undammed river, causing stress on fish and other organisms. The warmer water can also invite invasive species and parasites that introduce ecological and human health concerns.
- When dams flood vegetated land, particularly in tropical climates, the resulting decomposition can produce methane, potentially creating as much greenhouse gas during the years of decay as fossil fuel power plants of comparable output.

For these reasons, and because the damage incurred by the installation of large dams is difficult to reverse, many environmental organizations oppose further large-scale hydroelectric development. (See Figure 9.19.)

Since the late 1970s, the hydroelectric capacity in the United States has been increased largely by adding hydroelectric generators to existing dams built for other purposes (such as recreation, irrigation, and flood control), by upgrading older hydro plants to increase output, or by constructing smaller facilities that are less damaging to the environment.

FIGURE 9.19 The Hetch Hetchy River valley in California was dammed in 1923 to provide hydroelectric power and drinking water for the San Francisco Bay Area. Opposed by John Muir and the Sierra Club, the resulting reservoir dramatically altered the natural landscape. *Source:* Isaiah West Taber/The Sierra Club.

FIGURE 9.19 *(continued)* This more recent image of the Hetch Hetchy River Valley clearly illustrates the dramatic change in the natural landscape when compared with the image taken prior to 1923. Courtesy of Garry Hayes.

Chapter Summary

Hydropower has provided mechanical assistance since the founding of human civilization. Throughout history, hydropower has increased human productivity, reduced labor, fostered innovation and technological advancement, and founded some of our earliest cities. More potent and reliable than wind, hydropower is an ancient application of energy utilized around the world wherever suitable resources exist.

Whether applied to tumbling mountain streams, broad slow-moving rivers, or dammed reservoirs, hydropower offers numerous advantages over other forms of energy:

- Its operation consumes no fuel—the source of energy is free and renewable.
- Its operation releases no pollution and no greenhouse gases.
- It is extremely efficient at generating electricity.
- It has low operation and maintenance costs.
- It is very reliable.

- It offers the greatest operational flexibility of any power source, adjusting deftly to fluctuating electricity demand.

The primary environmental impact of hydropower comes from the construction of dams used for commercial power plants. Nevertheless, even these dams are not without benefit:

- Dams create reservoirs for recreation, drinking water, and irrigation needs.
- Dams store excess water during years of high precipitation to buffer years of low precipitation.
- Dams provide flood control, and some improve river navigation for commerce.

Commercial hydropower plays a vital role in the electric utility grid. It is the first to come on line after a power outage and the last to be shut down for maintenance. Nevertheless, few sites suitable for large-scale hydropower installations remain undeveloped, limiting future growth.

Review Questions

1. Many consider hydropower to be the first form of mechanical assistance utilized by early people. Why do you suppose hydropower predates other forms of renewable energy?

2. Compare small-scale hydropower with small-scale wind power. In what ways are they similar? In what ways are they different?

3. What advantages does hydropower have over photovoltaics? What disadvantages?

4. The gravitational potential energy of water is the source of hydropower. Nevertheless, water must flow across a turbine to produce electricity, and flowing water represents kinetic energy. Explain why considering only the kinetic energy of moving water is inadequate when determining the output of hydropower.

5. Consider two micro hydro installations on mountain streams. The first, located in a cold climate, provides electricity all year except in the winter due to ice. The second, located in a warm climate, offers electricity all year except in the summer due to dry conditions. For an off-grid residence, which situation is easiest to accommodate? Consider both the needs of the residence and alternate forms of available renewable energy.

6. Environmental groups generally oppose the construction of large dams for hydroelectric projects despite the promise of clean, renewable energy. What changes would be necessary for commercial hydropower to avoid the associated problems? Can this be done economically?

7. Micro hydro installations often obtain an overall efficiency of 60% or less. This is primarily due to friction within the penstock and associated piping. What can be done to minimize this loss and improve efficiency?

8. Would pumped storage be a viable option for micro hydro? Explain your reasoning.

9. Consider two commercial power plants located on different rivers. The first has a flow rate of 50,000 gallons per second and a head of 10 feet; the second has a flow rate of 500 gallons per second and a head of 1000 feet. Compare the amounts of electrical power these facilities can produce. Which type of turbine should each facility use?

Practice Problems

1. A micro hydro installation uses 100 feet of pipe to convey a flow of 1.9 gallons per second to a turbine located a vertical distance of 10 feet below the inlet. If the system produces 116 watts of power, determine its overall efficiency.

2. A commercial hydroelectric power plant uses pumped storage to increase its capacity for generating electricity during peak loads. If the power plant is 89% efficient at converting the potential energy of water into electricity and the pumping process is 86% efficient at lifting water from downstream back into the reservoir for reuse, determine the overall efficiency of producing electricity through pumped storage.

3. The occupants of a remote village rely exclusively on a single micro hydro facility for all of their electricity needs. The hydro plant diverts water from a mountain stream 1500 feet above the facility. The flow rate varies from a summer maximum of 30 gallons per second to a winter minimum of 5 gallons per second. Determine the power available (in kW) during the summer and winter, assuming the facility is 70% efficient.

4. A commercial hydroelectric power plant with an efficiency of 90% uses a large reservoir as a water source. When the reservoir is full, a flow rate of 2000 gallons per second generates 10 MW of electricity. After 5 years of drought, the same flow rate generates only 8.1 MW. By how much has the water level in the reservoir receded (in feet)?

5. A small barrier in a mountain stream diverts water into a pipe that descends to a hydroelectric facility located 615 feet below. A steel grate prevents large objects from entering the pipe,

but occasional storms can wash branches and debris over the grate, obstructing the flow. Suppose a heavy rainstorm decreases the hydroelectric output from 40 kW to 13 kW. Determine the reduction in flow rate (in gallons per second). Assume the hydroelectric facility has an efficiency of 72%.

6. An engineer designing a hydroelectric facility recalls that the flow rate (in gallons per second) is four times the value of the head (in feet) but has forgotten both numbers. If the facility is expected to produce 1.5 MW of power with an efficiency of 88%, determine both the flow rate and the head.

Biomass

Courtesy of Scott Grinnell.

Introduction

Biomass energy is as old as humankind, dating back to the first wood fires that provided warmth, light, and a place to gather. Biomass fueled the rise of civilization: wood fires cooked food, oil lamps banished darkness, and fermented grains gave rise to alcohol.

Biomass refers to all organic material created by the sun's energy through **photosynthesis**. It includes forest products such as firewood, wood chips, bark, and sawdust. It includes agricultural residue such as straw, corn stalks, and rice husks. It includes industrial wastes such as pulp and paper mill sludge, slaughterhouse grease, sugar cane pulp, and wood and textile scraps from manufacturing processes. It includes animal manure, sewage, and municipal garbage. It includes crops specifically grown to create energy, such as corn, soybeans, switchgrass, and algae.

Traditional forms of biomass—firewood, crop residue, and dried manure—have historically served as the principle sources of heating and cooking worldwide and continue to do so in many countries. Biomass energy in the form of grass and grains fueled humanity's early transportation system of horses, mules, and oxen that pulled carriages, hauled wagons, and dragged plows (Figure 10.1). Biomass energy in the form of tallow candles and animal oil lamps provided light and increased nighttime productivity. High-temperature manufacturing used charcoal, which burns hotter than regular wood. As late as the mid-1800s, the United States relied on wood and other biomass for almost all of its energy needs. However, the introduction of coal during the Industrial Revolution diminished the use of biomass, and today biomass accounts for only 4% of the total energy consumed in the United States.

FIGURE 10.1 People have long relied on horses and other animals to move and carry goods. Biomass—in the form of grass—fuels this form of transportation. This photo, taken in 1886, shows a pioneer family in Loup Valley, Nebraska. *Source:* National Archives, 69-N-13606C.

Photosynthesis

The process of photosynthesis in green plants absorbs carbon dioxide from the atmosphere, combines it with water, and uses the energy of sunlight to create carbohydrates and oxygen (Figure 10.2). These carbohydrates—which include various types of sugars, starches, and fibers—form the building blocks of all biomass material.

$$CO_2 + H_2O + Sunlight \rightarrow O_2 + Carbohydrates$$

The efficiency with which plants convert sunlight into biomass varies among species and conditions, with a maximum as high as 6% and an average closer to 1%. Despite this low efficiency, sunlight creates biomass at a rate exceeding the total world energy consumption by almost seven times. While only a small fraction of the world's biomass could be sustainably harvested, biomass nevertheless represents an enormous source of renewable energy.

The combustion of biomass to provide energy is the reverse process of photosynthesis: carbohydrates combine with oxygen to create carbon dioxide, water vapor, and heat. However, this process generates no more heat and creates no more carbon dioxide than would have occurred through natural decay processes. The amount of carbon dioxide added to the atmosphere during combustion is no more than the amount of carbon dioxide originally absorbed during photosynthesis. As a result, the use of biomass fuels does not contribute to global climate change like the burning of fossil fuels and is considered carbon-neutral.

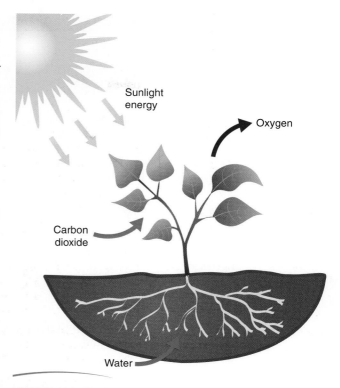

FIGURE 10.2 Photosynthesis converts sunlight into biomass by combining water and carbon dioxide to produce carbohydrates and oxygen. Carbohydrates form the building blocks of all biomass material. © 2016 Cengage Learning®.

Energy from Biomass

Different types of biomass provide different forms of useful energy. While some types of biomass can be burned directly to produce heat or generate electricity, others can be chemically or thermally converted into liquid or gaseous fuels. Four fundamental processes convert biomass into useful fuels, as shown in Figure 10.3.

- **Direct combustion** of dry biomass such as forest products, agricultural residue, and municipal solid waste can generate heat and electricity in much the same manner as conventional coal-fired power plants.
- **Biological conversion** of wet biomass such as livestock manure, municipal sewage, and agricultural silage can produce combustible gas through anaerobic digestion.

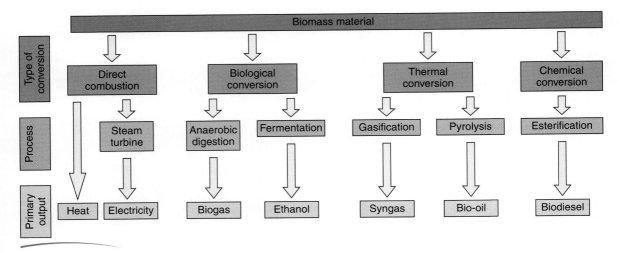

FIGURE 10.3 Four types of conversion processes generate useful energy from biomass. Depending on the process, the primary energy output can be heat, electricity, or various forms of gaseous or liquid fuels. © 2016 Cengage Learning®.

Grain slurry, sugar cane pulp, and processed cellulose can be fermented into ethanol as a source of liquid fuels.

- **Thermal conversion** of biomass in the absence of oxygen can produce combustible gas through gasification. Thermal conversion processes can also produce liquid fuels through pyrolysis.
- **Chemical conversion** of oilseed crops and microalgae can produce biodiesel.

Direct Combustion of Biomass

Direct combustion is the simplest and oldest use of biomass and continues to serve as the largest source of residential renewable energy. Wood-burning stoves and open fires provide warmth and a means of cooking throughout the world. The efficiency of burning wood to produce useful heat varies from as little as 10% for an open pit fire to as much as 75% for a sealed-combustion wood stove (Figure 10.4).

In regions with plentiful wood supplies, high-efficiency wood stoves offer an inexpensive and sustainable form of renewable energy. However, in many countries, the reliance on wood has led to severe shortages and caused substantial environmental damage through erosion, loss of soil nutrients, and ultimately desertification. Millions of acres of forest disappear every year, particularly in tropical regions, through unsustainable harvesting of firewood.

At the commercial scale, the direct combustion of biomass generates heat and electricity in much the same manner as burning coal (Figure 10.5). As early as the 1930s, paper and pulp factories began installing recovery boilers that concentrated pulpwood residue, extracted reusable chemicals, and generated heat and electricity for on-site power consumption. In the mid-1970s, utility-operated power plants began burning

FIGURE 10.4 High-efficiency wood stoves convert as much as 75% of biomass energy into usable heat. Courtesy of Scott Grinnell.

FIGURE 10.5 Wood chips serve as a renewable fuel source for the Bay Front Power Plant in Ashland, Wisconsin. Originally designed to burn coal, the plant added the capacity to use wood chips in 1979, and wood chips now supply more than 60% of the plant's 73-MW output. Courtesy of Scott Grinnell.

municipal solid waste, generating electricity from trash that would otherwise find its way into local landfills. The first utility-operated biomass power plant in the United States was established in 1984 in Burlington, Vermont, using wood chips to generate 50 MW of electricity.

Municipal Solid Waste Power Generation

Burning municipal solid waste in power plants is considered renewable energy because it consumes no new fuel—only material that would otherwise go to the landfill. Municipal solid waste consists of garbage collected through community sanitation services and includes household domestic waste; community waste from streets, parks, and playgrounds; and commercial waste from supermarkets and restaurants. Municipal solid waste consists primarily of paper, yard waste, food scraps, and wood products but also includes materials derived from fossil fuels such as plastic and rubber (Figure 10.6).

One of the primary benefits of burning municipal solid waste is the elimination of garbage that would otherwise take up space in landfills (Figure 10.7). Since the mid-1980s, more than three-quarters of all landfills in the United States have closed, reducing the total number of landfills from more than 8000 in the mid-1980s to about 1900 today.[1]

Landfills pose numerous environmental and human health concerns. As rainwater percolates through landfills, it leaches compounds from the decomposing garbage and can cause significant groundwater pollution. Although new landfills require impermeable liners to contain this liquid, studies estimate that more than 80% of these landfills continue to leak.[2] In addition to groundwater contamination, decomposing garbage in landfills produces methane, a greenhouse gas more potent than carbon dioxide. Methane is highly combustible, leading to possible fires and explosions if not

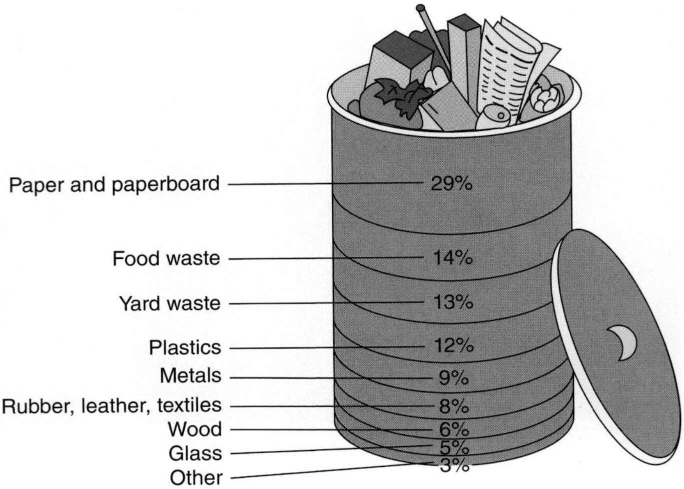

Paper and paperboard — 29%

Food waste — 14%

Yard waste — 13%

Plastics — 12%
Metals — 9%
Rubber, leather, textiles — 8%
Wood — 6%
Glass — 5%
Other — 3%

FIGURE 10.6 Most municipal solid waste generated in the United States consists of biomass that can be converted into useful energy. © 2016 Cengage Learning®.

FIGURE 10.7 The use of municipal solid waste for energy generation reduces demands on landfills and mitigates numerous environmental concerns. *Source:* David Parsons/NREL.

properly managed. Most new landfills capture and save the methane, which mitigates this problem (Figure 10.8).

Landfills attract vermin and scavenging birds, emit odors and asthma-inducing aerosols, and are generally unpopular when located near population centers. Consequently, most landfills are sited away from the communities that generate the trash and require long-distance trucking. Municipal solid waste power plants, on the other hand, typically reside within the population centers they serve, greatly reducing the expense and environmental concerns associated with trucking such as fuel consumption and air pollution.

FIGURE 10.8 The collection of methane from decomposing garbage at landfills reduces greenhouse gas emissions and provides valuable energy. © 2016 Cengage Learning®.

At the power plant, municipal solid waste is sorted and shredded, and specialized machines remove recyclable materials such as metals and glass. Agents such as lime and activated carbon may be added to promote clean burning. The processed waste then enters a combustion chamber, and the heat released through controlled burning drives steam turbines and generates electricity (Figure 10.9).

Municipal solid waste power plants serve the dual purpose of generating electricity while reducing demands on landfills. However, these power plants require careful design to minimize emitted pollutants. The sorting process cannot completely remove dangerous products like batteries and microelectronics that often contain toxic elements, including lead, mercury, cadmium, and rare earth metals. In addition, burning chlorine-containing substances found in many plastics can produce highly toxic and carcinogenic compounds called dioxins. Dioxins form in the combustion chamber or after combustion when gases cool in the exhaust stack. Municipal solid waste power plants can minimize the production of dioxins and other pollutants through high combustion temperature, adequate combustion time, and sufficient turbulence within the combustion chamber to evenly distribute heat through waste material. Also, the post combustion gases must be cooled quickly to minimize formation of dioxins and other pollutants within the hot flue gases. Electrostatic precipitators, filters, catalytic reactions, and other techniques further reduce pollutants.

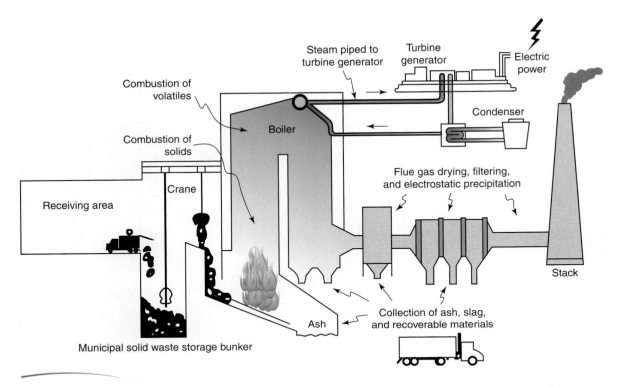

FIGURE 10.9 Municipal solid waste power plants burn shredded garbage to create steam that generates electricity in much the same manner as conventional coal-fired plants. Both types of power plants emit pollutants that must be controlled. © 2016 Cengage Learning®.

Crop Residue Power Generation

Traditionally, farmers burned crop residues or left them to rot on the fields—two methods that returned nutrients back to the soil. In the United States, roughly 80% of all agricultural crop residues come from the production of corn. The corn grain accounts for only 45% of the corn plant, and the remaining stalks, leaves, and husks (collectively called **stover**) are a valuable source of biomass (Figure 10.10). Today most corn stover is left on the fields after harvest to reduce erosion, improve soil moisture retention, and return organic material and nutrients back to the soil. Shredded corn stover also serves as animal bedding and mushroom compost. Nevertheless, several studies indicate that as much as 40% of corn stover could be sustainably removed and used as biomass feedstock as long as farmers rotate crops appropriately, sow cover crops, and minimize tillage.[3] Historically, the use of crop residues in high-temperature combustion chambers proved problematic due to the formation of **clinker**, a corrosive salt deposit resulting from the vaporization of potassium and chlorine present in the crops. While modern techniques can minimize this problem, gasification and fermentation processes more efficiently convert crop residues into useful energy.

FIGURE 10.10 Stover left over from harvesting corn can be baled and used for energy production. *Source:* Warren Gretz/NREL.

Biological Conversion of Biomass

Biological conversion of biomass utilizes microorganisms to break down organic matter into simpler constituents. Two types of biological conversions—anaerobic digestion and fermentation—create useful energy.

Anaerobic Digestion

Anaerobic digestion is a natural process in which bacteria biologically break down biomass in a low-oxygen environment to produce combustible gas, called **biogas**. Composed roughly of 60% methane and 40% carbon dioxide, biogas can be burned to generate heat and electricity or purified and concentrated and sold as natural gas. Anaerobic digestion is well suited for managing livestock manure, poultry litter, agricultural silage, municipal sewage, and other wet organic waste products.

Although biogas from anaerobic digestion may have been used in Assyria for heating bathwater as early as the tenth century B.C., the first documented anaerobic digestion plant dates back to 1859 at a leper colony in Bombay, India. In 1895,

FIGURE 10.11 Manure from animal feedlots often ends up in lagoons to decompose naturally. This process poses numerous environmental concerns, most of which are mitigated through anaerobic digestion. *Source:* Bob Nichols/USDA Natural Resources Conservation Service.

a sewage treatment facility in Exeter, England, began using biogas to fuel street lamps. Subsequent applications advanced the understanding of anaerobic digestion and increased the sophistication of equipment and operational technologies. Nevertheless, the Industrial Revolution largely halted interest in anaerobic digestion. Its revival coincided with the creation of large commercial feedlots, where managing livestock manure became expensive and environmentally problematic.

Livestock animals in the United States produce over 1 billion tons of manure each year, most of which is stored in outdoor lagoons and left to decompose naturally (Figure 10.11). This method of storage results in numerous problems:

- It potentially contaminates surface water and groundwater with excess nitrogen and other compounds.
- It causes air pollution by releasing ammonia, hydrogen sulfide, volatile organic compounds, and particulate matter—many of which cause health problems in humans.
- It releases greenhouse gases, including methane, nitrous oxide, and carbon dioxide.
- It attracts flies and other pests.
- It gives off unpleasant odors.

Anaerobic digestion mitigates the problems associated with storing manure in outdoor lagoons and generally reduces the overall cost of waste management. In addition, it kills pathogens and weed seeds in manure, producing a sterile and nearly odorless residue that can be used as fertilizer, mulch, and animal bedding.

Anaerobic digestion usually takes place in sealed containers, called digesters, that hold a slurry of organic waste and water. Three types of digesters are common:

- **Covered lagoon**: Manure-filled lagoons are covered with impermeable plastic and include manifolds to siphon off biogas as it forms. They are most applicable to dairy or swine operations in warm climates.
- **Complete mix**: An insulated tank made of steel or concrete heats manure to an optimal temperature for microbial digestion (Figure 10.12). More versatile than covered lagoons, complete mix digesters operate even in cold climates.
- **Plug flow**: Heated tanks with flexible covers allow for biogas collection. These are better suited to less liquid forms of dairy manure and can also be used in cold climates.

Modern sewage and wastewater treatment plants make use of anaerobic digestion as a means of reducing operating costs. The combustion of biogas released from decomposing sewage powers steam turbines to generate electricity and provides heat for on-site operations, saving millions of dollars in energy costs (Figure 10.13).

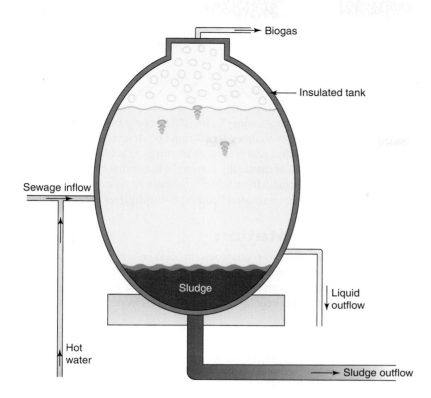

FIGURE 10.12 A complete mix anaerobic digester is an enclosed, insulated tank maintained at an optimal temperature to maximize microbial biogas production. © 2016 Cengage Learning®.

FIGURE 10.13 This anaerobic digester, consisting of 12 insulated tanks that are each 130 feet tall, is part of the Deer Island wastewater treatment plant in Boston, Massachusetts. Anaerobic digestion produces heat for the treatment plant and generates 28 MWh of energy each year. The remaining solids are processed into compost and fertilizer. © Greg Kushmerek/ www.Shutterstock.com.

Dairy farms are another industry well suited to anaerobic digestion. Each healthy dairy cow produces enough manure each day to generate about 2–3 kWh of electricity. The annual manure production per cow has an equivalent energy content of roughly 50 gallons of gasoline. Anaerobic digestion is a means of capturing this energy, and some diary farms are able to sell back more energy to the utility company than the farms require. However, dairy farms must be designed with efficient means to collect and transport manure to the digester. A common method of managing manure is to confine cows to feedlots or concrete tracts from which channels or pipes convey a mixture of manure and used bedding to the digester (Figure 10.14).

Fermentation

Fermentation is an anaerobic process in which yeast converts sugar into ethanol. Fermentation is one of the earliest biological reactions cultured by humanity, dating back at least 9000 years, and has been a source of alcoholic beverages throughout history.

$$\text{Sugar} \rightarrow \text{Ethanol} + \text{Carbon dioxide}$$
$$(C_6H_{12}O_6) \rightarrow 2(CH_3CH_2OH) + 2(CO_2)$$

FIGURE 10.14 Anaerobic digestion of animal manure requires a method to collect manure efficiently, such as confining animals to sheltered feedlots. © iStockphoto/36clicks.

Fermentation will not develop an alcohol content over 15% by volume, since this level becomes toxic for the yeast. Distillation is a process of evaporating and condensing ethanol to separate it from its surrounding fluid and increase its concentration. Distillation was well known by the time of the ancient Greeks and utilized throughout history. Today distillation techniques achieve ethanol concentrations of 95% by volume, and dehydration methods remove the remaining water to produce pure ethanol (Figure 10.15).

Distillation requires considerable heat input, supplied traditionally by burning crop residue or other available biomass. The energy required to create ethanol is important when considering the sustainability of creating ethanol as a fuel source. The overall efficiency of converting sunlight into biomass and then into ethanol fuel is quite low—much less than harvesting sunlight directly, such as with photovoltaic modules. However, when the biomass materials are waste products or agricultural residues or sustainably harvested surpluses, the production of ethanol becomes an important renewable energy source despite its low conversion efficiency. Furthermore, ethanol is valuable as a fuel for transportation, since liquid fuels contain much greater energy density than other forms, such as hydrogen or electricity stored in batteries.

In the United States, the production of ethanol fuel has been almost entirely through the fermentation of corn (Figure 10.16). Since 1980, the production of ethanol from corn has increased from 20 million gallons to more than 10 billion gallons.[4] Unfortunately, the use of corn—the edible grain—competes with food sources and is highly energy intensive. In fact, the corn industry consumes almost as much energy to produce ethanol

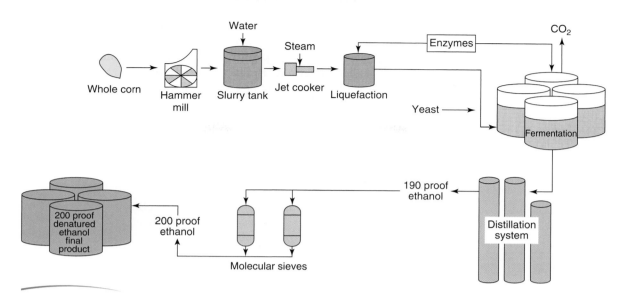

FIGURE 10.15 At an ethanol plant, corn is ground, mixed with water, mashed, and heated. Bacterial enzymes assist in the liquefaction of the mash, and yeast causes the mixture to ferment, producing ethanol and carbon dioxide. Energy-intensive distillation processes concentrate the ethanol up to 95%, and molecular sieves remove the remainder of the water, resulting in pure ethanol. © 2016 Cengage Learning®.

FIGURE 10.16 This ethanol plant, built in Nebraska in 1994, uses corn from nearby fields to produce 55 million gallons of ethanol per year. *Source:* Chris Standlee/NREL.

as the process yields. Conversely, ethanol produced from cellulose (the woody stock of the plant) rather than the grain does not compete with food sources and presents many opportunities. For example, the U.S. Department of Agriculture estimates that if 40% of corn stover is collected from farmers' fields and used as feedstock, it could produce 10 billion gallons of ethanol per year, an amount equal to the current grain-derived ethanol.[5] Other sources include crops grown specifically to serve as feedstock such as native grasses and fast-growing trees.

Further Learning ETHANOL IN TRANSPORTATION

The use of ethanol as a fuel for transportation has a long history. In 1826, ethanol powered the first prototype of an internal combustion engine. In 1908, when Henry Ford designed the first Model T automobile, he intended to use ethanol derived from renewable biomass. Ethanol presents many advantages over gasoline.

- Unlike gasoline, ethanol is nontoxic.
- Ethanol provides a higher octane rating, reducing engine knocking.
- Ethanol is less likely to explode or burn accidentally, making it safer in the event of an accident.
- Ethanol burns cleaner and releases fewer pollutants.
- Ethanol can be produced locally, whereas gasoline requires pipelines to distribute it from oilfields to population centers.

- Ethanol is simpler to manufacture, whereas gasoline requires complex refining procedures.

Despite the advantages of ethanol, gasoline dominated the liquid fuel market due to inexpensive supplies of petroleum, intense lobbying by petroleum companies, and a failure to consider environmental impacts. Until relatively recently, the enormous capital and well-established technology of the oil and automobile industries made the return of ethanol and other biofuels difficult. However, with increases in the cost of fossil fuels and in government incentives, ethanol produced from biomass has become economically promising even without considering its environmental and health benefits.

Thermal Conversion of Biomass

Thermal conversion is a process of heating combustible material with too little oxygen to allow burning. Instead, the applied heat drives off moisture, tars, and combustible gases, leaving behind a charred residue consisting almost entirely of carbon. The most familiar product of thermal conversion is ordinary charcoal, a form of renewable energy that has been in use for thousands of years (Figure 10.17).

Charcoal Production

People began producing charcoal as early as 7000 years ago for smelting copper, and its applications expanded during subsequent millennia to include all manner of high-temperature manufacturing, particularly metallurgy. Traditionally, people created charcoal by burying mounds of wood under damp earth and allowing the wood to smolder for a few days. The advantage of charcoal over wood is that charcoal is more easily transported, burns much hotter than wood, and contains almost no moisture, tars, or volatile gases that can contaminate manufacturing processes. However, charcoal retains only 20%–40% of the original energy of the wood, making it highly inefficient as a fuel. Furthermore, the smoldering process releases vaporized tars, oils, volatile gases, and toxic chemicals directly into the atmosphere. Finally, the gathering of wood for the production of charcoal has been one of the primary causes of deforestation throughout history and continues to be problematic in many countries.

Modern forms of thermal conversion are more sustainable and less damaging to the environment. These include gasification of biomass to produce combustible gas and pyrolysis of biomass to create liquid fuels.

FIGURE 10.17 Natural charcoal consists of the charred remains of wood. Although the production of charcoal is highly inefficient, it has been used for thousands of years as a fuel source for metallurgy and other high-temperature processes. Courtesy of Scott Grinnell.

Gasification

Gasification is a process that heats combustible material to high temperatures (1500°F–2000°F) in a vessel with too little oxygen for the material to burn. Instead of burning, gasification breaks down fuel into component molecules and creates a synthesis gas, or syngas, which consists primarily of hydrogen and carbon monoxide (Figure 10.18). By breaking fuel into molecules, gasification allows for the separation and capture of individual compounds. Hence, potential pollutants—including mercury, sulfur, and carbon dioxide—can be removed and reused. Since the vessel is fully contained, gasification produces no emissions.

Gasification offers a means of extracting energy from biomass and other burnable materials without generating the emissions characteristic of direct combustion. Furthermore, by breaking fuel into molecules, gasification has the potential to induce chemical reactions among components to create liquid fuels, fertilizers, industrial chemicals, and an array of consumer products.

The addition of steam to the gasification process converts carbon monoxide into carbon dioxide and additional hydrogen. The hydrogen-enriched syngas can directly power conventional gas-fired turbine generators to create electricity, or it can be purified and sold as hydrogen gas. Carbon dioxide can be captured and recycled or pumped underground for long-term storage. Therefore, when using biomass, which removes carbon dioxide

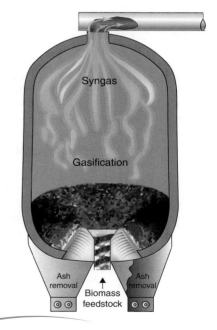

FIGURE 10.18 Combustion requires heat, fuel, and oxygen. With insufficient oxygen, heated biomass breaks down into volatile gases consisting primarily of hydrogen and carbon monoxide, called syngas. © 2016 Cengage Learning®.

FIGURE 10.19 In 1998, the McNeil Generating Station in Burlington, Vermont, began using wood chips in a gasification plant to produce 12–15 MW of electricity. This power supplements the original wood chip combustion facility, which has been operating since 1984. *Source:* Warren Gretz/NREL.

from the atmosphere through photosynthesis, the extraction and storage of carbon dioxide through gasification represents a carbon-negative process, reducing the overall carbon load of the atmosphere.

Although gasification of wood was used as early as the late 1800s to light street lamps, the first commercial-scale biomass gasification plant in the United States began operation in Burlington, Vermont, in 1998. Using local forest and lumber mill residue, the plant generates 12–15 MW of electricity (Figure 10.19).

Gasification power plants can use nearly any form of biomass, including forest and lumber mill products, crop residues, industrial wastes, sewage sludge, tires, asphalt roofing, and municipal solid waste. Furthermore, since gasification allows for the capture of carbon dioxide and other potential pollutants, it can be used to produce clean energy from coal. The combustion of coal in conventional power plants is the primary source of mercury, sulfur dioxide, and many other atmospheric pollutants. Gasification can minimize the impact of using coal and help ease the transition from fossil fuels to renewable energy.

Gasification is more versatile, more efficient, and cleaner than direct combustion, and its use will likely expand as technology matures and becomes more available.

Pyrolysis of Liquid Fuels

Pyrolysis is a process of heating combustible material in order to collect the volatile components and condense them into a liquid fuel, called bio-oil. Like gasification, pyrolysis takes place in sealed vessels with too little oxygen for the feedstock to burn. However, the process selects optimal parameters (including temperature, pressure, and oxygen content) to minimize gasification and promote the production of liquid fuels.

The most common method of producing liquid fuels is fast pyrolysis, a process in which moderately high temperatures (800°F–1000°F) act on the biomass for less than 2 seconds. The organic material rapidly decomposes into volatile gases, and prompt cooling condenses the gases into liquids (Figure 10.20).

To minimize contamination of the bio-oil with water, the moisture content of the feedstock must be less than 10%. The excess heat generated by the power plant is usually sufficient to dry arriving feedstock. If the feedstock is already dry, the excess heat can be used to provide local space or water heating. Since the biomass must volatize quickly, the

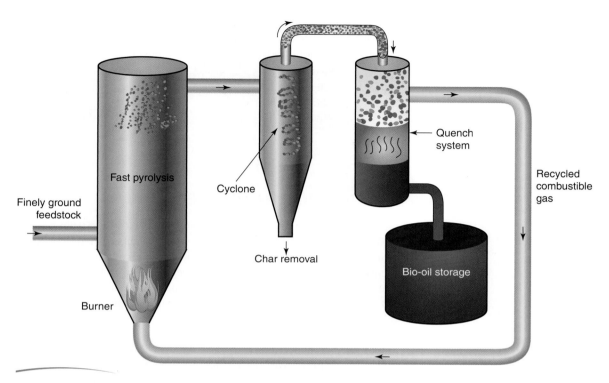

FIGURE 10.20 Fast pyrolysis quickly heats finely ground biomass to release volatile gases, separates out the solid char, and condenses gases into bio-oil, a liquid fuel much like crude oil. © 2016 Cengage Learning®.

feedstock must be finely ground, typically into pieces less than ¼ inch in diameter, and mechanisms must promote efficient char separation to prevent the char from interfering with incoming feedstock.

The nonvolatile remains, called **bio-char**, are largely composed of pure carbon and have a marketable value for use in manufacturing and consumer goods. Processing this carbon into fibers makes it available for marine, automotive, and aeronautical components; lightweight high-performance gear; sports equipment; toys; and many other products. Carbon can also be activated for use in fish tanks and odor filters.

Summary of the Thermal Conversion Processes

Each of the three processes of thermal conversion—charcoal production, gasification, and fast pyrolysis—generates different fractions of liquid fuel, char, and gas, as summarized in Table 10.1. The temperature and heating time vary among processes and result in different products.

TABLE 10.1 Thermal conversion processes.

PROCESS TYPE	DESIGN REQUIREMENTS		PRODUCTS GENERATED		
	Temperature	Heating Time	Liquid	Char	Gas
Charcoal production	Low	Long	30%	35%	35%
Gasification	High	Moderate	5%	10%	85%
Fast pyrolysis	Moderate	Very short	75%	12%	13%

© 2016 Cengage Learning®.

Chemical Conversion of Biomass

The **chemical conversion** of vegetable oil into biodiesel is a relatively simple process called **transesterification**. Although combustible, raw vegetable oil is too viscous to be used in conventional diesel engines and can leave undesirable combustion residues. Transesterification causes a chemical reaction between vegetable oil and an alcohol, creating biodiesel and glycerin. More than 85% of the original vegetable oil becomes biodiesel, and most of the remainder becomes glycerin. Once purified, glycerin can be used for soap, food additives, antifreeze, health care products, and many other manufactured goods.

Biodiesel may be used in conventional diesel engines without dilution or engine modification. Unlike petroleum-derived diesel, biodiesel is completely nontoxic, is biodegradable, and contains virtually no sulfur. Therefore, its combustion produces no sulfur dioxide, a component of smog and a source of acid rain. Biodiesel is also less explosive and hence safer in the event of a motor vehicle accident.

The production of biodiesel dates back to the mid-1800s—well before the widespread use of petroleum. Rudolf Diesel, the inventor of the diesel engine, originally designed his engines to operate on vegetable oils and biodiesel. At the time, he anticipated vegetable oil–based fuels powering farm equipment, allowing farmers to grow their own fuel. However, the inexpensive commercialization of petroleum eliminated most biofuel production until the 1970s, when the oil embargo and new environmental legislation encouraged alternative and cleaner fuel options.

In the United States, soybeans currently provide the primary feedstock for commercial biodiesel production, followed by palm oil from Southeast Asia. Oilseed crops—including canola, safflower, mustard, flax, sunflower, and rapeseed—can also serve as feedstock for biodiesel production (Figure 10.21). In addition, microalgae grown in large tanks or open ponds provide a source of oil. Oilseed crops contain between 20% and 45% oil by weight, and microalgae can approach an oil content of 50%.

Extracting biomass oil requires a series of steps. For seed crops, this includes separating the seed from the plant, crushing and flaking the seed, and then removing the oil with a screw press or expeller. Mechanical methods extract up to 80% of the oil, and further reduction using solvents can remove a total of 95%. The meal left over is highly nutritious and commonly used as livestock feed.

Like oilseed crops, microalgae can be mechanically pressed to extract the oil. Experimentation using ultrasonic vibration, which ruptures the cell walls of the algae,

FIGURE 10.21 Golden fields of rapeseed yield energy-rich seeds for oil production. Rapeseed, like many other oilseed crops, can provide feedstock for biodiesel. *Source:* Library of Congress, Prints & Photographs Division, photograph by Carol M. Highsmith, LC-DIG-highsm-18361.

FIGURE 10.22 Shallow ponds support the growth of microalgae. More productive than oilseed crops, microalgae promise to be an abundant source of oil for biodiesel production. *Source:* Nature Beta Technologies Ltd, Ei/NREL.

may provide an alternative method. Microalgae offer a promising feedstock for biodiesel production because of their high oil content, rapid growth, and large per-acre yield (Figure 10.22).

Whether using oilseeds or microalgae, the energy obtained from the crop must be greater than the energy used to prepare, plant, grow, harvest, and process the crop. The ratios of energy output to input depend on the farming practices and growing

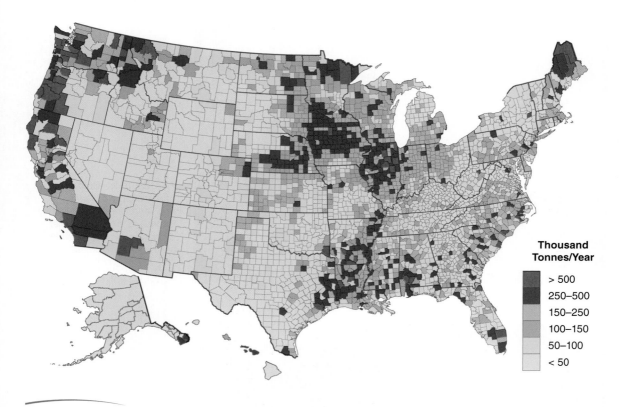

FIGURE 10.23 This biomass resource map, compiled by the National Renewable Energy Laboratory, includes waste in the form of animal manure and municipal sewage, methane emissions from landfills, and residues from agricultural crops and forestry products. © 2016 Cengage Learning®. *Source:* Adapted from Billy Roberts/National Renewable Energy Laboratory (NREL) data map.

conditions and can vary considerably. A map of the biomass resources of the United States, shown in Figure 10.23, includes waste in the form of animal manure and municipal sewage, methane emissions from landfills, and residues from agricultural crops and forestry products.

Dedicated Energy Crops

Dedicated energy crops are fast-growing trees, perennial grasses, and oilseed crops grown specifically to provide biomass for energy production. They include plantations of willow, poplar, and eucalyptus trees; prairie grasses like switchgrass, big bluestem, and miscanthus; and soybeans and other oilseed crops. Some of these, such as soybeans, serve as natural rotation crops for wheat and barley production to improve weed control and reduce chemical additive requirements.

The most useful energy crops are those that can be grown in ways that do not displace or compete with food crops, including those that can be grown on land unsuitable for food crops. A promising candidate is native perennial grasses (Figure 10.24), which once covered the prairies of the United States before settlers replaced them with annual food crops like corn and wheat. These grasses grow quickly and can be harvested year after year with minimal additives or soil amendments. They do not require tilling the soil—an energy-intensive process that contributes to erosion. Furthermore, these native grasses tolerate floods, droughts, pestilence, and poor soil conditions and can be harvested 10 years or more before replanting is needed. Unlike most conventional crops, the roots of prairie grasses extend below the surface nearly as far as the grasses extend above. As a result, they add organic matter to the ground even when harvested. Such grasses could also be used to slow storm water runoff, to anchor soil on stream banks and wetlands, and, when planted adjacent to conventional row crops, to reduce leaching of fertilizers, herbicides, and pesticides into nearby waterways.

FIGURE 10.24 Switchgrass, a fast-growing prairie grass native to North America, could serve as a dedicated energy crop for ethanol production.
Source: Warren Gretz/NREL.

Prairie grasses can be burned directly to produce heat and electricity or serve as feedstock for gasification or pyrolysis. Alternatively, they can provide feedstock for ethanol production or serve as a source of fiber for pulp and paper manufacturing. Although the production of ethanol from these grasses is currently expensive and not yet proven to be commercially viable, the technology is rapidly maturing. Prairie grasses may produce five times as much energy in the form of ethanol as required to grow them, a considerable improvement over corn.

Dedicated energy crops, when appropriately selected, offer environmental and economic benefits. They can:

- Improve water quality by reducing erosion and runoff.
- Increase native wildlife habitat.
- Enhance soil health by improving nutrient retention, carbon sequestration, and moisture control.
- Revitalize rural communities by providing local jobs and revenue.
- Provide a domestic carbon-neutral energy supply.

The most compelling application of dedicated energy crops is to create hydrogen or liquid fuels for transportation. The low conversion efficiency from sunlight into biomass through photosynthesis makes electricity production more practical from other forms of renewable energy, such as wind and solar.

Chapter Summary

Biomass includes a broad range of materials consisting of organic matter created by the sun's energy through photosynthesis. The efficiency of green plants in converting sunlight into biomass is quite low—only about 1% on average—so using biomass to generate electricity is far less efficient than more direct forms of renewable energy. Nevertheless, when biomass consists of sustainably harvested surpluses or wastes, it becomes a valuable source of renewable energy, particularly in the form of liquid fuels that contain the energy density crucial for transportation.

Four different conversion processes create energy from biomass: *direct combustion* of dry biomass like forest products, crop residue, and municipal solid waste to produce heat and electricity; *biological* conversion of wet biomass like manure, liquid sewage, and grain slurry to produce biogas and ethanol;

thermal conversion of dry biomass to produce syngas and bio-oil; and *chemical* conversion of oils to produce biodiesel.

The decomposition of biomass, whether through deliberate processes to produce energy or through natural decay processes, releases roughly equivalent amounts of heat and carbon dioxide into the atmosphere. Extracting energy from biomass does not necessarily deplete soil of essential nutrients, since many of the nutrients remain in the residue. The ash from combustion, the compost from anaerobic digestion, the seed pulp from oil extraction, and the grain mash from ethanol production contain the majority of the original nutrients. Although healthy soil requires organic material, an important fraction of biomass can be sustainably removed and utilized as long as these residues are returned to the soil.

Review Questions

1. Each year the United States generates 4 billion tons of trash, an amount equal to about 4 pounds per person per day. What do you think is the best method of dealing with municipal solid waste?

2. Explain why the burning of biomass is considered *carbon-neutral*.

3. While organic matter is important for soil health, the essential nutrients largely remain in the ashes and residues that are left after combustion. What happens to the stored biomass energy of organic matter if it is allowed to decompose naturally without being burned?

4. Burning crop residue has been a source of renewable energy for thousands of years. How does the addition of pesticides, herbicides, and other chemical additives during production influence the appropriateness of burning crop residue?

5. Different types of microorganisms act on different forms of biomass and yield different products. Distinguish between the two

biological conversion processes. What are the primary applications for the fuels resulting from these two processes?

6. Some products that are least able to be recycled, such as old asphalt roofing, provide energy-rich feedstock for biomass energy production. Name some other difficult-to-recycle products that share this feature. What form of biomass conversion would best utilize these products?

7. Anaerobic digestion from livestock manure typically requires animals to be confined to feedlots so that manure can be easily captured and utilized. What problems might this introduce?

8. Analysts claim that, in order for anaerobic digestion to be economical, dairy farms must have at least 400 cows. In India, individual families owning a few cattle use anaerobic digestion to produce gas for cooking. Explain the difference.

9. Both biological and thermal conversions of biomass require low-oxygen conditions.

For each of the processes, why is the absence of oxygen necessary? What products would result if oxygen was present?

10. Dairy farms—rather than swine, poultry, or beef cattle farms—comprise the vast majority of anaerobic digestion facilities in the United States. Why do you think this is so?

11. What is the primary benefit of gasification over direct combustion? Why is gasification not more widely utilized?

12. One of the benefits of fast pyrolysis is the creation of carbon, which can be processed into fibers and used to create rigid, strong, lightweight products. Compare the nonenergy benefits of pyrolysis with those of gasification. Which do you think is a better method of processing biomass?

13. The most common feedstocks for the production of biodiesel are locally grown soybean oil and palm oil imported from Southeast Asia. Which feedstock do you think has fewer harmful environmental impacts? Why?

14. Waste oils from restaurants, often available for free, can be an economical source for biodiesel production. When burned, the biodiesel emits odors characteristic of the foods cooked in the oil. How do you think this would impact the desirability of large-scale conversion of waste oil into biodiesel?

15. Two liquid fuels produced from biomass are ethanol and biodiesel. Both are nontoxic and biodegradable. Discuss the differences between the two fuel types.

16. The energy retrieved from burning or gasifying municipal solid waste is a small fraction of the embodied energy of creating the material in the first place. Although these processes mitigate problems associated with landfills, they are ultimately a poor substitute for reduced consumption and improved recycling. What methods should be implemented to reduce consumption and improve recycling?

17. The cultivation of dedicated energy crops is most useful only when they can be grown in ways that do not compete with or displace food crops. Describe a situation where this can be accomplished successfully.

Practice Problems

1. In a rural midwestern community, dry oak firewood costs $200/cord, while natural gas costs $1.65/therm. A cord of dry oak contains 30 million British thermal units (Btu) of energy, while each therm of natural gas equals 100,000 Btu. If the efficiency of a wood stove in converting biomass energy into useful heat is 75%, while that of a natural gas furnace is 92%, how much would it cost to heat a home requiring 50 million Btu using each fuel source? Which is more economical?

2. A dairy farm with 500 cows uses anaerobic digestion to produce 370,000 kWh of electricity per year by burning biogas in a gas-turbine generator. If the manure produced by each cow over the course of a year contains 6 million Btu of energy, what is the overall efficiency of converting manure into electricity? (1 kWh = 3412 Btu.)

3. Most ethanol currently produced in the United States comes from the edible corn grain. Each acre of corn produces about 350 gallons of ethanol. If ethanol was made from switchgrass instead, each acre could produce about 1100 gallons of ethanol. In addition, the energy required to grow and process switchgrass is half as much as that needed for corn.
   a. Compare the net energy output of producing ethanol from switchgrass to that of corn.
   b. If an acre of corn stover (rather than the grain) produces 180 gallons of ethanol and all the corn grain is used as food, how would this influence your choice of crop?

4. Mobile gasification units convert firewood into combustible fuel for automobiles. Invented during the 1930s, these units were common in Europe during World War II, when gasoline was scarce. Wood-powered automobiles travel roughly 1 mile per pound of wood scraps. For someone who purchases dry oak for $200/cord, what is the fuel economy in miles per

dollar? Each dry cord represents 3500 pounds. How does this compare to a gasoline vehicle that obtains 30 mpg?

5. Microalgae are a promising source of oil for biodiesel production. Proponents claim that algae can produce 50 times more oil per acre than soybeans.

   a. If each acre of soybeans produces 38 gallons of biodiesel each year, how much biodiesel could be made from an acre pond of algae?

   b. If the average person drives 10,000 miles per year in a vehicle obtaining 25 mpg, how many acres of algae would be necessary for each person?

   c. If the United States has 150 million people who drive vehicles, how many square miles of algae would be necessary to provide biodiesel for the entire country (1000 acres = 1.56 square miles), assuming the proponents' claims are true?

Endnotes

[1] U.S. Environmental Protection Agency. (2011). *Municipal solid waste generation, recycling, and disposal in the United States: Facts and figures for 2010* (EPA-530-F-11-005). Washington, DC: EPA.

[2] Environmental Research Foundation. (1992). *New evidence that all landfills leak*. Annapolis, MD: Rachel's Hazardous Waste News #316; G. F. Lee & A. Jones-Lee. (1992). *Detection of the failure of landfill liner systems*. El Macero, CA: G. Fred Lee & Associates.

[3] R.M. Cruise & C. G. Herndl. (2009). Balancing corn stover harvest for biofuels with soil and water conservation, *Journal of Soil and Water Conservation*, 64(4), 286–291; K.L. Kadam & J. D. McMillan. (2001). *Logistical aspects of using corn stover as a feedstock for bioethanol production*. Golden, CO: National Renewable Energy Laboratory.

[4, 5] U.S. Energy Information Administration. (2011). *Annual energy outlook 2011* (DOE/EIA-0383), Washington, D.C.: U.S. Department of Energy.

Non-solar Renewable Energy

Introduction

Nearly all energy available to humanity comes from the sun. Whether captured directly through photovoltaic modules or indirectly from biomass or fossil fuels or through the movement of wind or water, the sun powers our world. The two forms of energy not derived from sunlight—tidal energy and geothermal energy—are small by comparison. Nevertheless, these two non-solar forms of renewable energy offer valuable resources in locations where they exist. This chapter investigates tidal and geothermal energy sources and their importance to humanity.

Tidal Energy

Tidal energy results from the effects of the gravitational interaction of the moon and sun on the earth's oceans (see *Further Learning: Two Daily Tides* below). Ultimately, the source of tidal energy is the kinetic energy of the rotating earth. The daily motions of the tides dissipate 3.7 trillion watts of power and gradually slow the rotation of the earth, lengthening the day by about 20 millionths of a second every year.

Further Learning TWO DAILY TIDES

The gravitational attraction between the earth and the moon causes oceans to bulge outward on both the near and the far sides of the earth, resulting in two high tides and two low tides each day.

A perspective from earth suggests that the moon orbits around the earth. In actuality, the earth and the moon orbit about one another, each pulled by the same gravitational force felt by the other.

© 2016 Cengage Learning®.

This is similar to an adult and a young child holding onto a rope while ice-skating around each other. The child moves in a large circle around the adult and travels relatively quickly. The adult, however, skates in a tight circle, moving rather slowly. The same force of tension in the rope pulls each one toward the other, but the adult moves less quickly in response to this force due to a much larger mass. The earth and the moon comprise a similar system. At all times, the earth moves toward the moon, and as the moon circles around it, the earth constantly changes directions to follow the moon, like the adult facing the skating child.

The farther apart objects are located, the weaker the gravitational force acting between them. As a result, the gravitational force is strongest on the side of the earth closest to the moon, weaker at the earth's center, and weaker yet on the side most distant from the moon. Since the oceans are fluid and able to deform, they respond to the differing strengths of the gravitational forces by bulging in different directions. On the side closer to the moon, the oceans move toward the moon more than the earth itself because the force on the near side is greater than the force on the center of the earth. On the side farther from the moon, the oceans move toward the moon less than the earth itself because of the weaker gravitational force. The inertia of the oceans resists the change in the earth's motion much like loose clothing puffs outward whenever skating in a circle, resisting the change in the skater's motion. Similarly, the outward bulge on the far side of the earth is a result of the oceans being left behind as the earth adjusts its course to follow the moon.

The positions of the two tidal bulges remain nearly constant with respect to the moon while the earth rotates about its axis, causing the landmasses to move into the bulges and producing the observed tidal variations.

Although the moon provides the largest influence on the earth's tides, the sun is also important. When the gravitational forces of the sun and moon act on the earth in the same direction, such as during a full or new moon, the tides are a maximum, called **spring tides**. When their gravitational forces act on

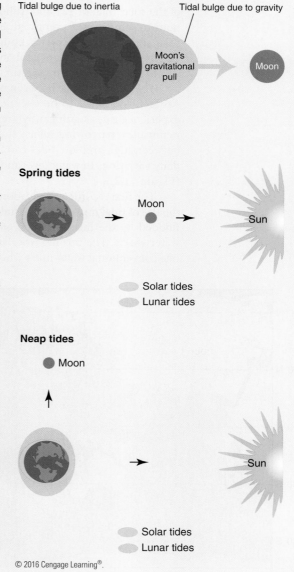

© 2016 Cengage Learning®.

the earth perpendicular to each other, such as during a first or last quarter moon, the tides are a minimum, called **neap tides**. The time between each spring and neap tide is one-quarter of the lunar month, or about 7.5 days.

While tidal power is minuscule compared with solar-derived forms (only 0.0030% of the solar total of 121,800 trillion watts at the surface), it offers substantial contributions to some coastal communities, particularly those located in areas that lack reliable solar sources. Tidal power is nonpolluting and highly predictable. It consumes no fuel, and its operation does not contribute to climate change. However, its application is limited to suitable locations, as estimated in Figure 11.1.

The average range between high and low tides in the middle of the ocean is about 1.6 feet, too small to be useful for power generation. However, many coastal sites experience tides significantly greater than this, amplified by topographic features like peninsulas, narrowing inlets, and shallow estuaries. Within coastal waters, the range between high and low tides can be as much as 50 feet or more, resulting in profound daily variations in water level and providing ample opportunities for extracting energy (Figure 11.2).

The first recorded applications of tidal energy date back a thousand years or more to the use of tidal mills along coastal regions of Spain, France, and England. Storage ponds flooded by the incoming tide emptied during low tide through channels fitted with waterwheels. The turning wheels provided mechanical power that ground grain much like conventional water mills. The first major tidal-electric power plant was constructed

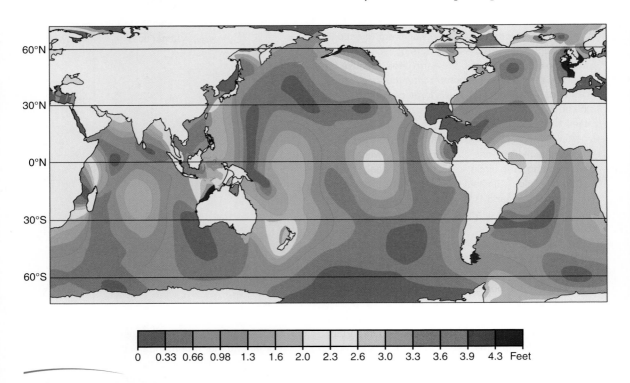

FIGURE 11.1 Locations that experience large tidal variations, shown in progressively deeper shades of red, allow some coastal communities to harness tidal energy. © 2016 Cengage Learning®.

FIGURE 11.2 Located in the Bay of Fundy, where tidal variations can be as large as 50 feet, Hall's Harbor in Nova Scotia experiences dramatic changes in sea level. Courtesy of Terri McCulock/Bay of Fundy Tourism, Nova Scotia.

near St. Malo, France, in 1966 and employed a similar technique, though it dammed an entire estuary to store water. The only tidal-electric power plant in North America operates in Nova Scotia, Canada, and was completed in 1984, making tidal electricity a relatively recent addition to world power.

Two fundamentally different techniques convert tidal energy into electricity: (1) the construction of shallow dams, called barrages, across inlets or estuaries offering high tidal range and (2) the deployment of underwater turbines that rotate in tidal currents.

Tidal Barrage

A **tidal barrage** generates electricity in much the same manner as a hydroelectric dam: a difference in water level across the barrage creates gravitational potential energy that can be utilized to rotate turbines (Figure 11.3). Unlike hydroelectric dams, however, tidal barrages can generate power by harnessing flow in both directions. As the incoming tide raises sea level on the ocean side, the passage of water through the turbines produces electricity and floods the estuary. When the tide ebbs and sea level is low, water trapped behind the barrage passes back through the turbines, again producing electricity. In this way, a tidal barrage behaves like a hydroelectric dam on a river that reverses direction four times a day.

Power Generation

As with a hydroelectric facility, power from a tidal barrage depends on two parameters: the flow rate and the head. The flow rate (Q) is the volume of seawater to pass through the turbines, measured in gallons per second (gal/s), and depends on the size of the turbines. The head (H), measured in feet (ft), is simply the difference in water level across the barrage. Therefore, the greater the range between high and low tides, the more power can be produced. The equation for tidal power (P) is the same as the equation for

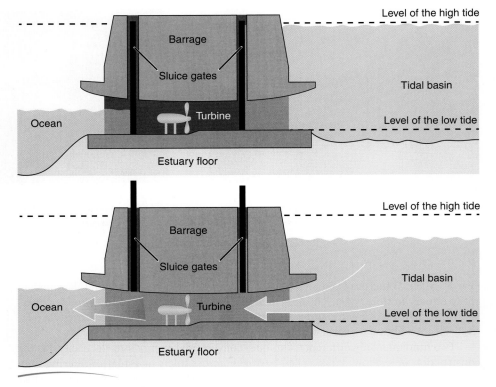

FIGURE 11.3 Turbines within tidal barrages act as low-head hydroelectric generators, sometimes producing power during both the incoming and the outgoing tides. © 2016 Cengage Learning®.

hydroelectric power, where g = gravity and ρ = density of seawater; however, seawater is slightly denser than freshwater:

$$P = \rho g Q H$$

$$P = 11.6 \frac{\text{watts}}{(\text{gal/s})\text{ft}} QH$$

Friction within the system and imperfections in turbine alignment and rotation speed reduce the actual power output. Since turbines perform optimally over a limited range of rotation speeds, most tidal barrages do not produce power continuously with the changing tides but delay the passage of water until a sufficiently large height difference, or head, exists across the barrage. When the head is adequate to allow efficient turbine operation, sluice gates open, and the power plant generates electricity. As the water levels equalize across the barrage and the head diminishes, the turbines lose efficiency. At high tide, the sluice gates close to capture as much water as possible, and as the tide approaches its minimum, the sluice gates reopen, and the stored water returns to the ocean, generating electricity a second time. Depending on the tidal range, a barrage may produce electricity only 3–4 hours during the roughly 6-hour tidal cycle (Figure 11.4). Although tides and the energy available from them are highly predictable, the generation of power is nonetheless intermittent and does not always align with energy demands.

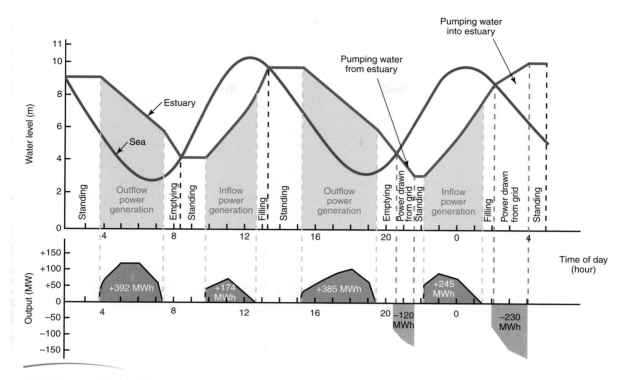

FIGURE 11.4 Differences in water level between the sea and estuary provide opportunities for power generation during both the incoming (flood) and the outgoing (ebb) tides. Operating the turbines as pumps consumes electricity, rather than creating it, but can make energy production more efficient overall and may help serve peak loads. © 2016 Cengage Learning®.

Source: Adapted from La Rance Tidal Power Plant, BHA Annual Conference Report, 2009.

Tidal barrages, like hydroelectric plants, have the option of augmenting the water level behind a barrage by actively pumping seawater into the estuary when the available head is too small to generate electricity. While this consumes energy rather than produces it, the additional water may allow the turbines to operate more efficiently during the next low tide. Furthermore, depending on the timing of energy demands, the additional capacity may help meet peak loads.

Environmental Impact

Tidal barrages impact their surroundings and the local ecology much like hydroelectric dams impact river ecology. Estuaries, which often present the greatest economic potential for tidal barrages, are among the richest, most diverse, and most delicate aquatic ecosystems. The obstruction of estuaries can be even more damaging than the damming of rivers. Tidal barrages present four primary concerns. The barrages:

- **Impede fish migration**: Many fish are born in freshwater, live out their lives in the ocean, and return to freshwater to spawn. These fish must pass through the turbines of an intervening barrage at least twice, and while the turbines are designed to allow fish passage, the mortality rate is significant.

- **Change the intertidal zone**: The area alternatively flooded and dry during the tidal cycles—known as the intertidal zone—is ecologically diverse and richly inhabited. Tidal barrages usually change the tidal cycles and water levels, potentially endangering this ecosystem.
- **Diminish natural flushing**: Barrages reduce the free flow of tides that provide a natural flushing of wastes and debris, leading to potential problems with water quality and silt accumulation.
- **Block navigation**: Although locks can be installed, they are slow and costly for boat traffic and deny free passage for marine mammals and other large aquatic life.

These concerns have been the basis for opposition from environmental groups and local inhabitants and have largely prevented further development of tidal barrages. High construction costs and a lack of suitable locations worldwide have further restricted development.

Underwater Turbines

An alternate method of harnessing tidal energy utilizes underwater turbines to intercept fast-moving tidal currents, or tidal streams. Rather than exploiting the potential energy created by a difference in water level, underwater turbines harness the kinetic energy of tidal streams. Similar in design to wind turbines, these devices can be deployed in shallow coastal waters and harness energy during both the flood and the ebb tides.

Tidal turbines offer notable advantages over wind power. Unlike the wind, tidal currents are highly predictable and produce energy at well-known intervals, facilitating easy integration into the electric utility grid. Since the density of water is 830 times greater than that of air, tidal turbines can be much smaller than wind turbines of similar output, reducing space requirements and the associated impacts. From the surface, tidal turbines are invisible and silent, mitigating the primary environmental concern of wind turbines: visual pollution.

Despite these advantages, the installation of any ocean energy devices—including tidal stream turbines—remains a fledgling industry. The corrosive and occasionally violent nature of the marine environment poses many technological challenges. Despite ongoing research and optimism, very few tidal stream turbines currently exist.

The two most promising designs for tidal stream turbines include (1) variable-pitch rotors mounted on underwater towers and (2) rotors enclosed within ducts and deployed on frames placed directly on the seafloor.

- Variable-pitch rotors maximize energy extraction by optimizing blade alignment. Mounted on towers like horizontal-axis wind turbines, they capture energy during both the flood and the ebb cycles either by pitching the blades completely around or by pivoting the entire turbine to face into the current (Figure 11.5). Variable pitch rotors have the advantage of being able to pitch the blades into a neutral position when necessary, stopping rotation without relying on mechanical brakes.
- Rotors enclosed within ducts and deployed directly on the seafloor mitigate concerns with ship traffic and disturb the seafloor less than turbines mounted on permanent towers (Figure 11.6a). Some designs employ flaring ducts that accelerate tidal flows through the turbines, improving performance (Figure 11.6b). In general, rotors enclosed within ducts operate less efficiently than variable-pitch rotors but are less costly, more quickly deployed, and more adaptable to multiuse locations.

FIGURE 11.5 Underwater turbines perform optimally when elevated above the seafloor. Towers supporting variable-pitch rotors can either be mounted on heavy stands, as shown in this artistic rendition, or anchored directly to the seafloor using semipermanent foundations. © 2016 Cengage Learning®.

FIGURE 11.6 (a) The Open-Centre Turbine, manufactured by OpenHydro, utilizes a hole in the center of a slowly rotating turbine to minimize injury to fish and other marine life. Courtesy of OpenHydro. (b) A design by Lunar Energy employs flaring ducts to concentrate currents and accelerate flow through the turbine. Courtesy of Lunar Energy.

Both types of turbines harness energy through the principle of lift, rather than drag, and convert kinetic energy into electricity with the same limiting efficiency of 59% as wind turbines. Turbulence caused by underwater obstacles and friction with the seafloor diminish the available energy. Mounting turbines on towers elevates them above the seafloor and closer to the surface, where tidal currents are strongest and least turbulent, maximizing energy production.

Although tidal turbine technology is still developing, advancements in wind turbine design and many years of hydroelectric experience promise rapid progress. Numerous countries around the world are investing in underwater turbine technology and anticipate deploying tidal energy devices in the near future.

Power Generation

Underwater turbines extract energy in the same manner as wind turbines, so the available power (P) depends on the same three parameters: (1) the speed of the current, v, measured in mph; (2) the area swept out by the rotor, $A = \pi r^2$, measured in square feet; and (3) the density of seawater, ρ.

$$P = \frac{1}{2}\rho A v^3$$

$$P = 4.25 \frac{\text{watts}}{\text{ft}^2\text{mph}^3} A v^3$$

The speed of tidal streams varies continuously during the ebb and flood cycles, causing the electrical output of tidal turbines to vary as well. Since the available power depends on the speed of the current cubed, underwater turbines produce most of their power during peak currents, when the tide is rushing in or out. Unlike with tidal barrages, which can regulate to some extent the release of captured water, the output of underwater turbines is less adjustable, producing an average of only 30% of their rated power.

EXAMPLE 11.1

Estimate the electricity produced from an underwater turbine of rotor area 215 square feet for a tidal stream flowing at a peak speed of 9 mph during its 6-hour cycle. Estimate the changing flow speed as follows: 9 mph for 1 hour, 6 mph for 1 hour, 4 mph for 1 hour, 2 mph for 1 hour, and close to zero for the remaining 2 hours.

Solution: The total energy generated over the 6 hours equals the sum of the energy produced at each speed: $E_{TOTAL} = E_1 + E_2 + E_3 + E_4$, where energy equals power multiplied by time (in watts $\times$ hours). Each kWh equals 1000 watt-hours.

$$
\begin{aligned}
E_{TOTAL} &= 4.25 A v_1^3 T_1 + 4.25 A v_2^3 T_2 + 4.25 A v_3^3 T_3 + 4.25 A v_4^3 T_4 \\
&= 4.25(215 \text{ ft}^2)(9 \text{ mph})^3(1 \text{ hour}) + 4.25(215 \text{ ft}^2)(6 \text{ mph})^3(1 \text{ hour}) \\
&\quad + 4.25(215 \text{ ft}^2)(4 \text{ mph})^3(1 \text{ hour}) + 4.25(215 \text{ ft}^2)(2 \text{ mph})^3(1 \text{ hour}) \\
&= 666 \text{ kWh} + 197 \text{ kWh} + 58 \text{ kWh} + 7 \text{ kWh} \\
&= 928 \text{ kWh}
\end{aligned}
$$

More than two-thirds of the total energy produced by the turbines occurs during the single hour of peak speed. By comparison, the turbine produces very little electricity during the final hour of tidal flow.

Environmental Impacts

Underwater tidal turbines may have one of the smallest environmental impacts of any existing energy production method. Nearly all fish and marine mammals that inhabit rapidly flowing tidal waters are sufficiently agile and perceptive to avoid slowly rotating turbines, and the turbines have minimal impact on other sea life. Unlike tidal barrages, underwater turbines offer a means of harnessing tidal energy without significant environmental concerns.

Underwater turbines located near the surface may interfere with shipping traffic, fishing, and recreational uses, while those placed on the seafloor offer unobtrusive deployment but compromise efficiency.

Further Learning THE WORLD'S FIRST COMMERCIAL TIDAL POWER PLANTS

I. TIDAL BARRAGE

The world's first tidal power plant was constructed near St. Malo, France, between 1961 and 1966. It consists of 24 10-MW low-head turbines within a tidal barrage that extends almost half a mile across the inlet of La Rance estuary. The power plant generates up to 240 MW, produces an average of 500 million kWh of energy each year, and has been in continuous operation since its creation.

La Rance estuary spreads over more than 8.5 square miles and holds roughly 50 billion gallons of water. It experiences an average tidal range of 27 feet (with a maximum of over 44 feet) and a maximum flow rate through its inlet of almost 5 million gallons per second. The low-head turbines, each measuring 17.5 feet in diameter, generate their rated power of 10 MW with a head of 18.5 feet and a flow rate of 73,000 gallons per second.

The construction of the tidal barrage involved creating two temporary dams that spanned most of the inlet, severely restricting flow into the estuary and precipitating a partial collapse of its ecosystem. Many forms of marine life, both flora and fauna, disappeared during this time. With the completion of the barrage, the temporary dams were removed, and the daily exchange of water across the barrage gradually restored the ecosystem. Proponents claim that, after more than 40 years of continuous operation, La Rance estuary remains a rich and diverse ecosystem. They contend that appropriate design and management, suitably timed flow through the barrage, and an understanding of the ecology at risk allow tidal barrages to be a source of ecologically responsible renewable energy. Others argue that complex and often poorly understood interactions take place within estuaries, making the risks unwarranted and the money better spent on conservation or other forms of energy generation.

II. TIDAL CURRENTS

The first commercial underwater power-generating facility to capture tidal currents, called *SeaGen,* was installed in Northern

© 2016 Cengage Learning®.

(continued)

(continued)

Ireland's Strangford Lough in 2008. Consisting of two 52-foot-diameter variable-pitch rotors, the facility generates up to 1.2 MW of power and delivers about 400,000 kWh of energy into the utility grid each year.

The two turbines rotate slowly—less than 15 revolutions per minute—and attach to a central tower that protrudes above the water's surface, allowing the turbines to be lifted completely out of the water for maintenance.

Strangford Lough is an inlet sheltered from the open ocean. It is an environmentally sensitive area, a breeding ground of seals, and home to porpoises and other large marine mammals. Conscientious monitoring of *SeaGen* assesses both the feasibility of underwater turbine technology and its environmental impacts. Since beginning operation, the environmental impacts have thus far proven minimal, and marine mammals apparently maneuver around the turbines without injury.

Geothermal Energy

Geothermal energy comes from the earth's interior. The immense temperature difference between the earth's core (at about 9000°F) and the cooler exterior results in enormous heat flow toward the surface (Figure 11.7). This geothermal heat drives convective motions within the earth, fragments the earth's crust into tectonic plates, and moves the plates around the surface at a rate of about an inch per year. It also generates volcanoes and earthquakes, pulls apart ocean basins, forces up mountain ranges, and occasionally heats groundwater to create geysers, fumaroles, and hot springs.

Throughout history indigenous people all over the world used hot water from geothermal sources for cooking, cleaning, bathing, and healing. Dwellings built near hot springs took advantage of the naturally occurring warmth. Romans built public bathhouses using geothermal heat for medicinal and leisure applications. Early Polynesian settlers in New Zealand relied on geothermal heat for cooking and heating. All major hot springs in the United States have historical significance with Native American tribes, and archaeological evidence of their use dates back more than 10,000 years.

Modern applications of geothermal energy began with spas and resorts in the early 1800s. The first production of electricity occurred in 1904 in Larderello, Italy, where a small steam generator produced enough power to illuminate five light bulbs. The first commercial geothermal power plant began operation in 1911 at the same location (Figure 11.8) and remained the only commercial geothermal power plant until 1958, when a second facility, Wairakei Power Station, began operation in New Zealand. In 1960, California installed the first commercial geothermal power plant in the United States, and California continues to dominate the country with 43 of the 69 geothermal installations.

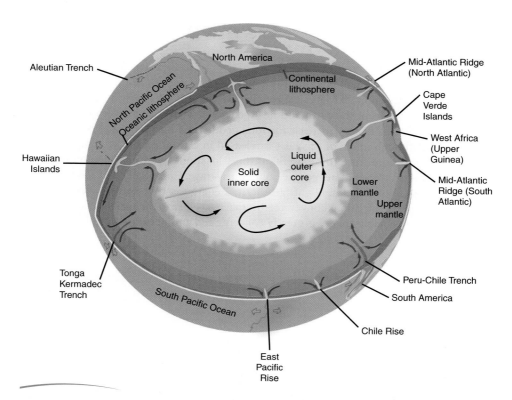

FIGURE 11.7 The transfer of heat from the earth's hot interior to the cooler surface occurs largely through slow convective motions within the mantle. Where hot plumes penetrate close to the surface, geothermal energy gives rise to naturally occurring geysers and hot springs and offers a means of generating electricity. © 2016 Cengage Learning®.

FIGURE 11.8 Today the geothermal facilities of Larderello generate over 700 MW of power and average almost 5 billion kWh of electricity per year. Steam emerges from the ground at 400°F, drives turbines to produce electricity, and cools through evaporation in the towers shown.

© Drimi/www.Shutterstock.com.

Source and Extraction of Geothermal Heat

The primary source of geothermal energy is the decay of radioactive elements within the earth. Small quantities of long-lived radioactive isotopes, principally uranium 238 and thorium 232, give off roughly 30 trillion watts of heat. Primordial heat from the formation of the earth and heat generated by the decay of shorter-lived isotopes early in earth's history increase the total geothermal heat output to roughly 44 trillion watts. Although much of this heat is released in crustal rocks, it has generally been too deep and too diffuse to be useful at most locations, warming the continental surface by a mere 0.065 watt per square meter on average. Compared with the 1000 watts per square meter of direct sunlight, geothermal energy is tiny. Nevertheless, in locations where magma intrusions convey geothermal heat close to the surface (as shown in Figure 11.9), the available energy is greatly concentrated and becomes useful for both direct heating

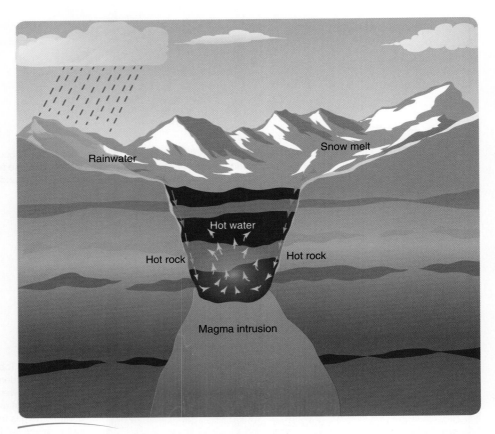

FIGURE 11.9 Intrusions of magma bring geothermal heat close to the surface. Groundwater seeping through fractures may penetrate deep enough to encounter high-temperature rock. The heated water may return to the surface naturally or be harnessed by drilling boreholes. © 2016 Cengage Learning®.

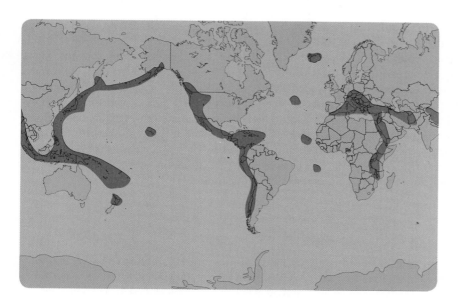

FIGURE 11.10 Only in locations where magma penetrates close to the surface (shown in red) is geothermal energy easily accessible. These locations generally coincide with tectonic boundaries where volcanoes, earthquakes, and hot springs naturally occur.

© 2016 Cengage Learning®. *Source:* Adapted from Geothermal Education Office Geo Presentation, 2000.

and electricity production. Figure 11.10 identifies known regions of concentrated geo-thermal resources around the world.

The rate of geothermal energy production within the earth is more than twice the rate of energy consumed by all human-related activities. However, only a small fraction of the world's geothermal energy could ever be harnessed, making it a valuable form of renewable energy—but one of limited capacity.

Although excavations of deep mines during the sixteenth and seventeenth centuries led people to deduce that ground temperature increased with depth, it was not until much later that scientific instruments documented the roughly 80°F average increase in temperature per mile of depth. Since most borehole-drilling machines penetrate only to a depth of 2 miles, the maximum temperatures accessible at most locations have gener-ally been too low to be useful for power production. Only recently have advances in drilling technology (derived from the petroleum industry) allowed deeper boreholes and greater geothermal opportunities. As a result, relatively few locations worldwide utilize geothermal energy for electricity generation, and those that do lie within concen-trated heat zones adjacent to tectonic boundaries where available temperatures typically exceed 270°F.

EXAMPLE 11.2

A mine penetrates half a mile below the surface. If the temperature at the surface is 60°F, what temperature would be expected at the bottom of the mine?

Solution: The increase in temperature would be roughly 80°F/mile × 0.5 mile = 40°F, raising the temperature at the bottom of the mine to 60°F + 40°F = 100°F.

Modern geothermal power plants extract energy by circulating fluid through geothermally heated rocks. This fluid is usually a solution of water containing relatively high concentrations of dissolved minerals and salts leached from the hot rocks. Although most early power plants simply dumped this mineral-rich brine on the surface, environmental concerns over surface-water contamination, depletion of underground aquifers, and possible settling of the ground prompted new power plants to reinject extracted fluids. In addition, while geothermal energy is a renewable resource on a global scale, it can be locally depleted and must be monitored to prevent overuse. The three oldest sites—Larderello, Wairakei, and the Geysers (in California)—all have reduced production from their peaks due to overextraction. Pumping additional fluid into the extraction zones can sometimes revitalize production.

Further Learning GROUND-SOURCE HEAT PUMPS

Ground-source heat pumps, often confused with geothermal energy, are unrelated to the production of heat within the earth. Despite commonly being called *geothermal heat pumps,* these systems owe their energy to a different source entirely: the sun. The sun delivers to the earth's surface approximately 1000 watts per square meter, a quantity vastly greater than the continental geothermal contribution of 0.065 watt per square meter. The ground serves as an enormous solar collector, storing energy like a thermal battery.

A ground-source heat pump is a system for heating (or cooling) a building by transferring heat from (or to) the ground. The system uses an electric pump to circulate fluid through pipes—usually made of plastic tubing—buried in the ground or submerged in water. Ground-source heat pumps take advantage

of the relatively stable temperature of the ground to draw heat *from the ground* during the winter when the ground is warmer than the air and to dump heat *into the ground* in the summer when the ground is cooler than the air. An electric compressor amplifies temperature differences and creates an efficient method of heating or cooling buildings, generally transferring three to five times more energy than the electricity required to operate the system.

Ground-source heat pumps take advantage of the relatively stable temperature of the ground to boost efficiency and reduce heating and cooling costs. The greater the depth, the smaller the seasonal variability in ground temperature. At a depth of approximately 20 feet, the ground remains at a temperature approximately equal to the average annual surface temperature.

Power Generation

The principle for generating electricity from geothermal energy is similar to that for conventional steam power plants, except that the heat of the earth, rather than the combustion of fossil fuels, provides the steam. With any steam-driven turbine, the hotter the steam, the more efficiently it can be converted into electricity. Since most geothermal sources produce steam that is considerably cooler than the temperature within conventional fossil fuel boilers (only 300°F–430°F compared with 900°F–1000°F), the efficiency of geothermal power plants is less, ranging from 10% to 20% compared with the 35% typical of fossil fuel plants.

The most effective method of utilizing geothermal energy depends on its temperature and whether the reservoir consists of steam or hot water. Three different technologies convert geothermal energy into electricity:

- **Dry steam power plants** are the simplest and oldest design, making direct use of geothermal steam as it emerges from the ground (Figure 11.11a). For the power plant to be operated economically, the steam must be at least 350°F and be free of liquid water. Only two large-scale power plants in the world utilize dry steam: the Geysers (Figure 11.12) and Larderello (Figure 11.8). The efficiency of dry steam power plants in converting geothermal heat into electricity is roughly 18%–20%.
- **Flash steam power plants** are the most common type in operation today (Figure 11.11b). These power plants tap into reservoirs of hot water rather than steam. Although the temperature of the hot water is typically 350°F or greater, the enormous subterranean

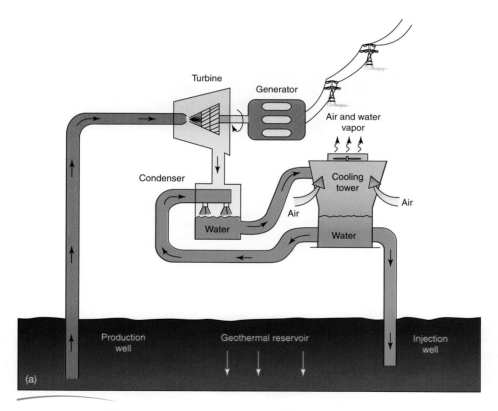

FIGURE 11.11 (a) When the geothermal resource consists of high-temperature steam without the presence of liquid water (called dry steam), it may be used directly to spin turbines and generate electricity. As steam emerges from the turbine, condensers cool it into liquid water to reduce the backpressure on the turbine and increase the turbine's efficiency. The hot water emerging from the condenser either passes through cooling towers, where evaporation creates clouds of visible steam, or passes through air-cooled radiators using fans, producing no emissions whatsoever. Some of the cooled water flows back to operate the condenser, while the remainder is reinjected into the reservoir to replenish the aquifer. © 2016 Cengage Learning®.

FIGURE 11.11 (b) When the geothermal resource consists of high-temperature water at high pressure, the water quickly vaporizes into steam when the pressure is released at the power plant. This steam drives turbines and passes through condensers, cools in evaporation towers or fan-driven radiators, and then returns to the aquifer through injection wells along with any remaining geothermal fluid. © 2016 Cengage Learning®.

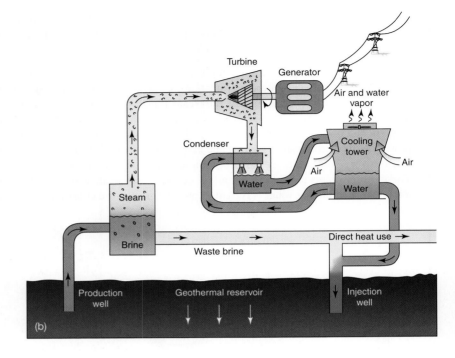

FIGURE 11.11 (c) Instead of piping geothermal fluid through turbines directly, a binary cycle power plant transfers geothermal heat to an operating fluid that boils at a much lower temperature than water. The operating fluid vaporizes in the heat exchanger, drives the turbine, and converts back into liquid in the condenser. The geothermal fluid returns to the aquifer, forming a closed loop that minimizes atmospheric emissions. © 2016 Cengage Learning®.

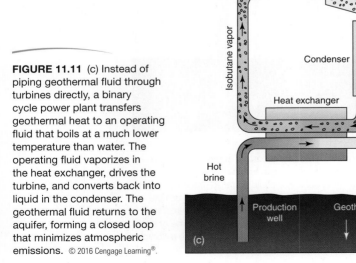

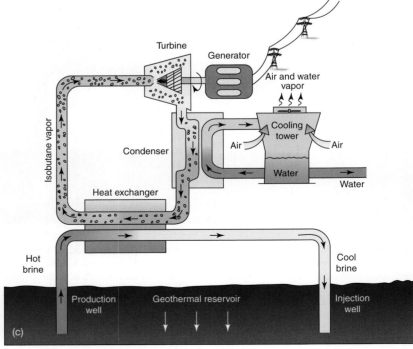

pressure prevents the water from boiling until it arrives at the surface and the pressure is released, at which time some of the water quickly vaporizes into steam and drives turbines. The remaining water, as well as the condensed steam, returns to the geothermal reservoir through reinjection wells, replenishing the aquifer and preventing surface contamination (Figure 11.13). These power plants also operate at efficiencies of 18%–20%.

- **Binary cycle power plants** employ a relatively new technology that provides a means of utilizing geothermal sources with temperatures less than 300°F (Figure 11.11c). The hot geothermal fluid passes through a heat exchanger and boils a secondary fluid that vaporizes at a much lower temperature than water. The vaporized secondary fluid (such as pentane or isobutane) drives turbines in much the same way as steam. This is the most common type of power plant being considered for future developments, since it can make use of lower-temperature resources that are much more abundant and widely distributed. Furthermore, since the geothermal fluid moves in a closed loop, it releases virtually no pollutants into the atmosphere (Figure 11.14). These power plants operate with an efficiency of only 10%–13% due to the cooler temperatures.

FIGURE 11.12 The first commercial geothermal power plant in the United States began operation in 1960 at the Geysers in northern California and makes use of the largest known dry steam reservoir in the world.
During 50 years of continuous operation, the Geysers geothermal area supported as many as 24 power plants and produced a peak output of almost 1600 MW. Currently, the Geysers operates 18 units and generates a combined output of about 750 MW. Over its 50-year history, the Geysers has delivered 280 billion kWh of electricity onto the utility grid (an average of 5.6 billion kWh per year), most of which was used within the San Francisco Bay Area.
Source: David Parsons/NREL.

FIGURE 11.13 Located in central California, the Coso Hot Springs geothermal facility consists of nine power plants using flash steam technology. Beginning commercial operation in 1987, the facility has produced as much as 2.3 billion kWh of electricity per year, though its current output has declined from a peak generation of 270 MW to less than 200 MW today.
Source: J.L. Renner/NREL.

FIGURE 11.14 The Mammoth-Pacific facility consists of three power plants built between 1984 and 1990 and generates 29 MW through binary cycle technology. Located near the Mammoth Lakes ski resort, the facility incorporates an air-cooling system to avoid visible plumes of steam. In addition, the facility consists of low-lying structures painted green to blend into the natural setting. After vaporizing isobutane gas in a heat exchanger, the cooled geothermal fluid is reinjected into the ground, creating a self-contained and emission-free system. *Source:* J.L. Renner/NREL.

Direct Heating

Geothermal reservoirs with temperatures less than 300°F—and others unsuited to the production of electricity—offer a source of direct heating for homes, offices, and greenhouses. Furthermore, low-temperature reservoirs around the world provide useful heat for aquaculture, food-processing plants, and a variety of industrial applications.

Most direct heat geothermal systems utilize wells drilled into reservoirs that provide steady streams of hot water. Since geothermal water is laden with dissolved minerals and noxious gases, a heat exchanger transfers heat to freshwater before circulating it through a network of heat-delivery pipes. The cooled geothermal fluid emerging from the heat exchanger is then injected back into the ground to prevent surface contamination and to replenish the reservoir.

In addition, waste heat from electricity production at existing geothermal power plants could be adapted to supply direct heat to communities and industrial centers whenever the plants are located within a few miles of the point of use. Conversely, advances in binary cycle technology may allow some low-temperature reservoirs to generate electricity in addition to providing direct heat.

Environmental Impacts

Fluids circulating through hot rocks contain a variety of dissolved minerals, salts, and gases that pose risks if released into the environment. Noxious gases (including hydrogen sulfide, ammonia, and sulfur dioxide) that escaped from early geothermal projects

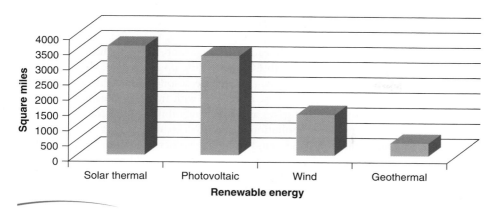

FIGURE 11.15 A comparison of the required physical space to generate 1 million kWh of energy per year reveals that geothermal power plants disturb less total area than other forms of renewable energy. Land requirements take into account service roads, a central station to convert solar thermal energy into electricity, the maximum shaded area of photovoltaic arrays, and the foundation platforms for wind turbines. © 2016 Cengage Learning®. *Source:* Data from Geothermal Energy Association.

drew criticism from environmental organizations and deterred the construction of some facilities. However, the techniques of reinjecting geothermal fluids and chemically trapping noncondensable gases (such as carbon dioxide and hydrogen sulfide) remedy most of these issues and allow geothermal power plants to operate nearly emission-free. In addition, geothermal power plants require less space and disturb less land than conventional power plants of similar output, as shown in Figure 11.15.

Although some environmental damage occurs when drilling boreholes, particularly if drilling fluid is not properly managed, traditional geothermal facilities can operate with remarkably little environmental impact.

Emerging Technology

The generation of electricity from the three types of geothermal power plants (dry steam, flash steam, and binary cycle) requires the presence of near-surface reservoirs of steam or hot water, severely restricting viable locations. Recent technology developed by the oil and gas industry has drastically expanded the potential for geothermal energy extraction. This technology is applied to create an **enhanced geothermal system (EGS)**, which allows the extraction of geothermal energy from relatively cool (less than 300°F) dry crustal rock as deep as 5 miles beneath the surface, making geothermal energy promising at most locations worldwide.

EGS relies on techniques of **hydraulic fracturing**, or **fracking**, in which water (usually mixed with sand and corrosive chemicals) is injected deep underground at high pressure to create new fractures in the rock and widen existing fractures. Hydraulic fracturing enhances the permeability of the rock formation, allowing water to move freely through a system of interconnecting fissures. Water injected into the newly created reservoir absorbs heat as it percolates through the hot rock formation, eventually finding its way to one or more strategically located extraction wells. The extraction wells transport the geothermally heated water back to the surface, where conventional binary cycle turbines generate electricity. As in other binary cycle power plants, the mineral-laden brine is reinjected underground and reheated, creating a closed-loop, nearly pollution-free cycle. However, the deep wells associated with EGS may bring up radioactive elements, most notably radon, that might escape into the atmosphere despite condensing and trapping techniques. Other concerns include those associated with fracking in general: groundwater contamination, migration of gases and hazardous fluids to the surface, induced seismic activity, and air and water pollution related to deep-well drilling.

Proponents anticipate that EGS could increase geothermal electricity production in the United States by more than 40 times its current level, adding up to 100 billion watts of electricity to the national energy grid and satisfying more than 10% of the total U.S. electricity demand. Others speculate that exploiting even a small fraction of the available geothermal reserves in the continental United States could provide more than 2000 times the total current U.S. electricity needs. Whether EGS delivers on its promise depends on further development, funding, and additional research. Figure 11.16 identifies locations within the United States that offer favorable EGS potential, as well as current geothermal sites.

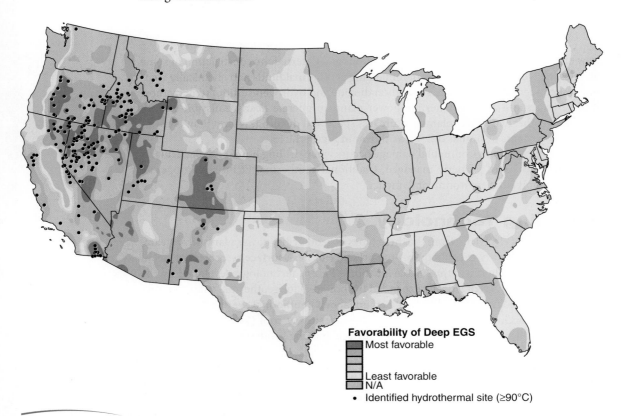

Favorability of Deep EGS
Most favorable

Least favorable
N/A
• Identified hydrothermal site (≥90°C)

FIGURE 11.16 In the United States, the most favorable geothermal resources—those with accessible temperatures above 300°F—are in California, Nevada, Oregon, and Idaho. © 2016 Cengage Learning®.

The promise of EGS appears to be vast, particularly as a means of producing clean energy to transition away from fossil fuels. However, the projected levels of geothermal extraction generally exceed the rate of heat generation within the crust, sometimes by as much as a thousand times. The global consequences of extracting heat from the earth faster than natural processes replenish it has so far not been adequately addressed or fully considered by proponents of the technology.

The first commercial success for EGS in the United States occurred in April 2013 at the Desert Peak geothermal field in Nevada. Funded by the U.S. Department of Energy, the project revitalized an unproductive geothermal well using EGS technology, increasing the well's output by nearly 38% and producing an additional 1.7 MW of electricity.

Chapter Summary

The two forms of renewable energy not derived from sunlight—tidal and geothermal—provide limited sources of renewable energy. While small compared to solar-derived forms, tidal and geothermal energy can serve as important clean energy sources for many locations around the world.

The planetary motions of the moon and the sun produce tidal variations that are highly predictable. Nevertheless, tidal power is intermittent and does not always align with energy demands. The environmental concerns associated with tidal barrages make them less promising than underwater turbines, which can be invisible and silent and offer one of the smallest environmental impacts of any energy generation system.

Geothermal power plants provide continuous and reliable electricity that is immune to fluctuations in weather and planetary cycles. This reliability gives geothermal a distinct advantage over solar, wind, and tidal energy. Although the efficiency of converting geothermal heat into electricity is less than that of conventional steam power plants, it consumes no fuel, produces almost no pollution, and generates electricity with fewer interruptions than most other methods. However, while geothermal energy is renewable on a global scale, it can be depleted locally if overexploited, particularly if geothermal fluids are not reinjected. Advancements in deep-well drilling, enhanced geothermal systems, and low-temperature binary cycle technology promise to make geothermal energy more widely available.

Review Questions

1. The type of waterwheels once used in tidal pools to grind grain resembles conventional waterwheels used on streams. What advantages and disadvantages would tidal mills have over stream-powered mills?

2. Underwater turbines provide a means of harnessing energy from any type of marine current. What is the primary difference between tidal currents and ocean currents (discussed in Chapter 8 in *Further Learning: Ocean Current Power*)? How does this difference influence turbine design?

3. Which harnesses more tidal energy, a tidal barrage spanning the inlet of an estuary or a farm of underwater turbines spread across the seafloor of an identical inlet? Why?

4. As underwater turbines extract energy from tidal currents, they reduce the speed of the current. What effect could this have on the ecology of coastal inlets and estuaries? Should this be considered when assessing the environmental impacts?

5. Geothermal energy is generated within the earth largely through the decay of radioactive elements. Which of the four fundamental forms of energy (see Chapter 5) is ultimately the source of geothermal energy?

6. Geothermal energy provides more than 80% of the residential heating and hot water requirements for the country of Iceland. Why does geothermal not provide a similar proportion of heating and hot water requirements in the United States?

7. People often confuse ground-sourced heat pumps with geothermal energy. Ground-sourced heat pumps circulate water through underground pipes—sometimes buried no more than 8 feet below the surface—to extract heat from the ground to warm buildings. What is actually the primary source of thermal energy in the earth's upper surface?

8. Explain why a *single* geothermal power plant, rated at 2-MW capacity, might produce the same amount of energy over the course of a year as *five* wind turbines, each also rated at 2-MW capacity.

9. A fumarole is a vent from which steam and volcanic gases escape into the atmosphere.

A geyser is an intermittently erupting spring of boiling water. If these two natural geothermal phenomena were harnessed to generate electricity, which type of power plant would each represent?

10. Suppose a revolutionary technology engineered a method to harness *all* of the earth's geothermal energy in a nonpolluting manner. What consequences would this have for the earth?

11. Suppose a natural hot spring is used to heat a public bathhouse. The overflow from the bathhouse is cleansed of any bathing impurities and returned to the original spring. Describe the environmental impacts, if any, of utilizing this resource.

12. Which form of renewable energy would most likely be able to supply all of humanity's energy needs single-handedly? How might this be done with minimal environmental impact?

Practice Problems

1. At a particular coastal site, a pool captures 500 million gallons of water at high tide and releases this water during a 30-minute period at low tide, when the vertical drop between the pool and ocean is 20 feet.
   a. Estimate the maximum power production (in watts).
   b. If the average power over the 30 minutes is half the maximum power, what is the total energy produced during a tide (in kWh)?

2. A commercial tidal current power plant, located in the Strangford Narrows of Northern Ireland, consists of two turbines, each with a rotor that measures 52 feet in diameter. The company claims that each turbine produces 600 kW of power in currents moving at 5.4 mph.
   a. Determine the turbine's efficiency in converting tidal current kinetic energy into electricity.
   b. If each turbine produces 5000 kWh of electricity per tidal cycle, what is its average power production (in kW)? Each tidal cycle lasts 12 hours and 25 minutes.
   c. Estimate the annual energy produced from both turbines (in kWh), assuming two equal tides per day.

3. A tower-mounted underwater turbine and a similar vertical-axis wind turbine both have rotors that measure 32 feet in diameter. If both harness turbulent-free flows of 18 mph, compare the power generated by each (in kW). Consider the difference in fluid density: $\rho_{SEAWATER} = 830\rho_{AIR}$.

4. A wind turbine with a 450-foot-diameter rotor produces 3.3 MW of power at a rated wind speed of 20 mph. What rotor diameter would allow an underwater turbine to generate the same power with tidal currents of 6 mph? Assume both turbines operate at the same efficiency. Consider the difference in fluid density: $\rho_{SEAWATER} = 830\rho_{AIR}$.

5. Suppose a geothermal power plant and a fossil fuel power plant both generate 10 MW of electricity. The geothermal power plant uses steam at 300°F and operates at an efficiency of 15%. The fossil fuel power plant uses steam at 1000°F and operates at a higher efficiency of 32%.
   a. Compare the rates at which each type of power plant produces waste heat.
   b. Which type of power plant offers better opportunities for supplying direct heat to residential and commercial buildings?

6. A ground-sourced heat pump circulates water through boreholes drilled to a depth of 1/10 mile (528 feet).
   a. If the average surface temperature is 45°F, estimate the temperature at a depth of 1/10 mile.
   b. What is the average temperature of the borehole pipe between the surface and its lowest point of 1/10 mile?
   c. If the circulating water begins at the surface temperature of 45°F and warms to the average temperature, by how much does its temperature increase?
   d. If each gallon of water absorbs 8.35 British thermal units (Btu) of heat for every 1°F increase in temperature, how many gallons of water must circulate through the ground-sourced heat pump to provide a home with 260,000 Btu of heat per day?

7. At a geothermal site, drilling a 2-mile-deep borehole costs $2 million and provides access to water at 350°F. Alternatively, drilling a 3-mile-deep borehole costs $4 million and provides access to 420°F water. The efficiency of generating electricity is 17% with 350°F water and 20% with 420°F water.
   a. If a geothermal power plant could produce 10 MW of power using 350°F water from the 2-mile-deep well, what output (in MW) would be expected from the 3-mile-deep well, assuming other conditions are equal?

   b. How much additional energy (in kWh) would be produced each year by the deeper well if the facility generates power 85% of the time?
   c. If electricity costs $.12/kWh, how long would it take to pay off the additional cost of drilling a deeper well?

8. A 20-MW geothermal power plant serving an urban community can be built in two locations. At the first site, located 20 miles away from the community, a 3-mile-deep well accessing 650°F water could be drilled for $8 million. A similar well drilled at the edge of the community would cost $20 million but would also provide hot water for district heating.
   a. If the power plant operates at 25% efficiency, how much waste heat (in MW) does it generate?
   b. Calculate the thermal energy generated each year by waste heat (in kWh) if the facility produces power 85% of the time.
   c. If the net value of district heating is $0.03 per kWh, determine annual savings to the community by utilizing the power plant's waste heat.
   d. If the cost of installing the infrastructure for district heating is an additional $20 million, how many years of operation would be required to recover the expenses of siting the plant at the edge of the community?

Glossary

A

active solar hot water system A type of solar hot water system that uses an electric pump to move fluid.

active solar system A solar heating or cooling system that relies on mechanical assistance, such as blower fans or pumps, to transfer heat within a building.

adobe A building material consisting of sand, clay, and a fibrous material like straw that is formed into blocks and dried in the sun.

air infiltration The amount of air to leak around the frame or between the seals of a window. Measured in cubic feet per minute per square inch of window surface (cfm/in^2); smaller values indicate less air infiltration.

alternating current (AC) A current that reverses direction many times a second, causing electrons to oscillate back and forth, propagating energy through the conductor without any of the electrons traveling the distance.

alternative construction Traditional building techniques using local, readily available materials.

altitude The maximum height of the sun above the horizon. This height, occurring at solar noon, changes seasonally.

amorphous solar cell A solar cell in which noncrystallized silicon is deposited in a very thin film directly onto a substrate material.

anaerobic digestion A natural biological conversion process in which bacteria biologically break down biomass in a low-oxygen environment to produce combustible gas known as biogas.

aperture An opening, such as a window, by which solar energy enters the interior of a building.

asphalt shingle A common type of roofing composed of paper or fiberglass saturated in asphalt and coated with ceramic granules.

awning mount system A mounting system that attaches collectors to a vertical wall using horizontal brackets adjusted in length to achieve the desired tilt.

awning window A type of window that opens upward to the outside.

B

bamboo flooring Made from a renewable, fast-growing grass common to tropical latitudes, bamboo flooring is composed of laminated strips.

batt or roll insulation Insulation in the form of flexible, blanketlike slabs made to fit into cavities between building studs.

binary cycle power plant A type of geothermal power plant in which a hot geothermal fluid (less than 300°F) passes through a heat exchanger to vaporize a secondary fluid that drives the turbines.

bio-char The nonvolatile remains of thermal conversion processes composed primarily of pure carbon.

bio-oil A liquid fuel created through pyrolysis that consists of acidic, tarry compounds.

biogas A combustible gas, composed roughly of 60% methane and 40% carbon dioxide, that can be burned to generate heat and electricity or purified and concentrated and sold as natural gas.

biological conversion A process of biomass conversion that utilizes microorganisms to break down organic matter into simpler constituents. Two types of biological conversion are anaerobic digestion and fermentation.

biomass Organic material created by the sun's energy through photosynthesis.

biomass energy A form of solar energy created through photosynthesis in living organisms that includes timber and agricultural products, sewage and livestock manure, and municipal garbage.

brick and stone siding A type of siding composed of brick or natural stone that is usually supported by the building foundation.

building envelope   The roof, floor, and exterior walls of a structure, including windows and doors, that separate the interior spaces from the outdoor environment.

C

carbon-negative   A process that reduces the overall carbon content of the atmosphere.

carbon-neutral   A process that over its life cycle neither adds carbon to the atmosphere nor removes carbon from the atmosphere.

carpets   Carpets are thick, woven floor coverings. Carpets vary substantially in composition and manufacture, resulting in considerable variation in durability, comfort, and healthfulness.

casement window   A type of window hinged on one side that cranks or pushes open.

cavitation   The sudden formation and collapse of bubbles that cause damaging turbulence.

cedar shake   A traditional wooden roof shingle made by splitting cedar logs.

cellulose   The woody stock of plants.

cellulose insulation   Insulation made from recycled paper (primarily newsprint) that is shredded into small, fibrous pieces and coated with a fire retardant.

cementitious foam insulation   A type of spray-in-place foam composed of magnesium oxide that dries to be fireproof, nontoxic, and resistant to water and mold.

ceramic tile flooring   Made of kiln-baked clay, ceramic tile has a fairly high embodied energy but provides long-lasting, low-maintenance, fireproof flooring.

CFL   Compact fluorescent bulbs produce light with only a quarter of the energy consumed by incandescent bulbs.

check valve   A one-way valve used in some solar hot water systems to prevent fluid from flowing backwards.

chemical conversion   A process in which biomass, such as oilseed crops and microalgae, can produce biodiesel.

chemical energy   The energy associated with the arrangement of electrons in atoms or molecules. Chemical energy is released whenever atoms or molecules establish or break bonds or alter their arrangement with each other.

clay tile   A roofing shingle made from clay that entails conditioning, shaping, and kiln baking.

clinker   The incombustible residue, often in the form of lumps or nodules, that is left over from burning coal.

closed-loop pressurized system   An active, closed-loop solar hot water system that utilizes a small, variable-speed electric pump to circulate a mixture of water and antifreeze through a collector. A heat exchanger transfers heat from the solar fluid to the domestic water.

closed-loop solar hot water system   A type of solar hot water system that circulates solar fluid through collectors and transfers heat to the domestic water with a heat exchanger.

cob   A building material consisting of sand, clay, and straw molded into a single, seamless entity by working layers together.

cold climate design   A well-insulated, compact building design appropriate for cold climates in which the organizing principle is to minimize heat loss.

comfortable   One of the six green building principles that includes constant temperature and humidity, adequate lighting, suitable acoustics, and a sense of place.

commercial hydroelectric power plant   A hydroelectric facility with a capacity that varies from less than 1000 kW to more than a million kW.

complete mix   A type of anaerobic digester consisting of an insulated tank made of steel or concrete that heats manure to optimal temperatures and allows for biogas collection.

compression seal   Elastic weather-stripping between the window sash and the frame that compresses as the window is tightened against it and minimizes air leakage around the sash.

concrete floor   A type of flooring that serves as both the foundation of the building and the finished floor.

conduction   The transfer of heat by direct contact.

conductor   A material, such as metal, that allows electrons to move freely.

control mechanism   A method or mechanism to regulate the amount of solar energy that enters a building, such as awnings or roof eaves.

convection   The transfer of heat through the motion of a fluid, such as air or water.

cordwood   A method of construction in which short logs are arranged transversely like a stack of firewood and mortared together.

cork flooring   Composed of the bark from the Mediterranean cork oak tree, this type of flooring offers long-lasting, shock-absorbing, and sound-deadening properties.

cotton insulation   Insulation consisting of recycled cotton (such as blue jean manufacturing trim waste) that is shredded, coated with a fire retardant, and pressed into batts.

covered lagoon   A type of anaerobic digester in which a manure-filled lagoon is covered in plastic and strategically placed manifolds siphon off biogas as it forms.

current   The flow of electric charge per unit of time, such as the number of electrons passing through a wire each second.

cut-in speed   The minimum wind speed capable of turning the rotor of a wind turbine and generating power.

cut-out speed   The maximum safe wind speed, above which a wind turbine ceases to operate.

D

density of air (ρ)   A measure of the mass of air per unit of volume. Air density varies with elevation, temperature, and humidity, being the greatest at low elevation under cold, dry conditions.

desertification   The degradation of fertile land into barren, desertlike conditions caused by mismanagement of resources, climate change, or other factors.

digester   A sealed container in which anaerobic digestion takes place for the purpose of creating biogas.

dimensional lumber   Lumber milled and cut to standard sizes and used throughout the building industry.

direct combustion   A process of burning dry biomass, such as wood chips, to generate heat and electricity.

direct current (DC)   A steady, continuous flow of electrons through a circuit, with each electron traveling the entire length of the circuit.

direct gain system   A passive solar design using south-facing windows, appropriate controls, and heat-absorbing surfaces with adequate thermal mass to directly warm interior spaces with sunlight during the day.

direct heating   A process of piping hot water from geothermal reservoirs into buildings as a method of providing heat.

distillation   A process of evaporating and condensing ethanol to increase its concentration.

domestic water   The potable water used within a household.

doping   A process of deliberately diffusing minute quantities of carefully selected impurities into a very pure crystal.

double-hung window   A type of window that allows individual sashes to slide up and down past each other.

drag   A force acting parallel to the direction of wind flow caused by a difference in pressure between the front and back surfaces.

drag-based design   A method of capturing wind energy with surfaces oriented perpendicular to the wind direction.

drainback system   An active, closed-loop solar hot water system in which the solar fluid drains out of the collector and back into a small reservoir when not in use. A heat exchanger transfers heat from the solar fluid to the domestic water.

drainback tank   A small reservoir that stores solar fluid for use in a drainback system.

dry steam power plant   A type of geothermal power plant that makes direct use of steam as it emerges from the ground. The steam must be at least 350°F and free of liquid water.

dual-axis tracking   A tracking system that follows the path of the sun in both the east-to-west direction and the vertical direction.

E

earthbag   A method of construction in which sacks filled with an inorganic earthen material are stacked on top of each other.

earthship   A type of rammed-earth tire structure promoted by Michael Reynolds in the 1970s.

Earthship philosophy   A philosophy that promotes simplicity, self-reliance, and the use of natural energy sources, such as solar and wind.

efficiency   The ratio of the output energy to the input energy.

electrical energy   The energy associated with the positions and motions of charged particles. Electrical energy manifests itself in three common forms: electricity, chemical energy, and electromagnetic energy.

electrical power   The rate of using electrical energy, equal to the product of voltage and current.

electricity   The organized motion of free electrons through a material.

electromagnetic energy   The energy associated with electric and magnetic fields produced by the motion of electric charges. Visible light, microwaves, and radio signals are common forms of electromagnetic energy.

embodied energy   The energy required to mine, harvest, manufacture, transport, and install a particular material.

energy   A quantity equal to the power multiplied by time, usually expressed in kilowatt-hours (kWh).

energy efficient   One of the six green building principles that includes the embodied energy of creating a building, as well as the operational energy of maintaining the building.

energy recovery ventilator (ERV)   A mechanical device that provides ventilation while exchanging heat and moisture between stale exhausted air and fresh intake air.

engineered wood   Any building material created by binding together wood material (veneers, strands, flakes, or particles) with adhesives to create a composite manufactured wood product.

engineered wood flooring   A flooring material composed of three or more layers of wood veneer glued together.

engineered wood siding   A siding material composed of sawdust mixed with an adhesive binding agent and made to resemble natural wood.

enhanced geothermal system (EGS)   A system for generating geothermal electricity by enhancing the natural permeability of deep rocks through hydraulic fracturing.

environmental degradation   A measure of the damage caused to the environment, including erosion, pollution, and ecological destruction.

environmental impacts   Avoiding harmful environmental impacts is one of the six green building principles and includes minimizing pollution, erosion, and habitat destruction.

equinoxes   The dates when the sun's rays are perpendicular to the equator, on or about March 20 and September 22, producing the same length of day every place on earth.

ethanol   A colorless, flammable, clear liquid formed by microbial fermentation of carbohydrates that can be used as a fuel in internal combustion engines.

evacuated tube collector   A type of solar hot water collector consisting of numerous sealed evacuated tubes of annealed glass that individually attach to an upper manifold.

expanded polystyrene (EPS) insulation   Closed-cell foam insulation made from polystyrene beads expanded and fused together. EPS foam can be molded into various forms for packaging or made into rigid sheets for construction.

extruded polystyrene (XPS) insulation   Closed-cell foam insulation made from polystyrene beads that are melted and extruded into rigid boards that are usually colored by the manufacturer to designate brand.

F

fast pyrolysis   A thermal conversion process that briefly heats combustible material to 800°F–1000°F and condenses the volatile components into a liquid fuel, called bio-oil.

fermentation   A biological conversion process in which yeast converts sugar into ethanol.

fiber-cement   A composite building material composed of sand and Portland cement surrounding a fiber matrix that is commonly used for siding or roofing.

fiber-cement siding   A type of siding made from Portland cement, sand, and cellulose fiber that is often textured and painted to resemble natural wood.

fiber-cement tile   A roofing shingle made primarily of Portland cement over a fiber matrix.

fiberboard   A structural board created from pulpwood that is mechanically broken down into fibers, felted, and then reconstructed into sheets by heat and pressure.

fiberglass insulation   Insulation made from molten glass that is spun or blown into fine fibers. It is available as rolls, batts, or loose fill.

First Law of Thermodynamics   Energy is neither created nor destroyed. It is simply converted from one form to another.

fixed mount system   A nontracking mounting system that holds an array at a set angle.

flash steam power plant   A type of geothermal power plant that generates electricity by causing hot water (350°F or greater) to vaporize rapidly into steam at the surface.

flat plate collector   A type of solar hot water collector consisting of a shallow insulated box enclosing fluid-carrying tubes attached to absorber plates and protected behind tempered glass.

flow rate (Q)   The volume of water that arrives at a hydroelectric turbine or waterwheel each second, measured in *gallons per second*.

fossil fuels   A form of solar energy derived from prehistoric algae, bacteria, and plants that were buried and compacted under layers of sediment.

foundation   The portion of a building that supports and distributes the structure's load into the ground.

fracking   A common term for hydraulic fracturing.

frame quality   A measure of the structural and insulating properties of a window frame. A high-quality window frame not only supports panes of glass but also includes a thermal break to minimize conduction.

framing   The use of uniform structural elements, such as standardized dimensional lumber, to erect a stable frame for attaching interior and exterior wall coverings.

Francis turbine   A reaction turbine in which the water flows through a curved tube around the turbine and guide vanes direct the water radially inward to strike the turbine blades.

friction   The resistance to flow. For hydroelectric facilities, friction reduces the water pressure at the turbine and lessens its effective head.

furling   A method of limiting excessive rotation rates in residential-scale wind turbines by tilting or rotating the turbine out of the wind stream.

G

gasification   A thermal conversion process that heats combustible material to high temperatures (1500°F–2000°F) to produce combustible gas called syngas.

geothermal energy   A nonsolar form of energy resulting primarily from the decay of radioactive elements within the earth.

geothermal reservoir   A region of geothermally heated rocks or aquifer capable of providing useful energy.

gliding window   A type of window that allows one or more sashes to slide past each other horizontally.

gravitational potential energy   The energy associated with the positions of objects within the presence of gravity.

green building   A high-performance building that minimizes resource consumption and harmful impacts on the environment, while producing healthy living spaces for people.

greenhouse gases   Gases that absorb and emit infrared radiation in the atmosphere and contribute to the warming of the earth's surface.

Griggs-Putnam Index   An index used in site assessments that links long-term prevailing wind patterns with coniferous tree deformation.

ground mount system   A mounting system that supports solar collectors on a ground-anchored frame that tilts the collectors to optimize performance.

H

hardboard   A thin, stiff board of compressed sawdust and wood fibers.

head (H)   The height from which water descends, measured in *feet*, which determines the pressure at the turbine.

healthful   One of the six green building principles in which structures promote good health through the absence of toxins, poisons, allergens, and other harmful substances.

heat-absorbing surface   A surface on the interior of a building that absorbs solar energy after it has passed through the aperture.

heat exchanger   A device for transferring heat between the solar fluid and domestic water.

heat recovery ventilator (HRV)   A mechanical device that provides ventilation while exchanging heat between stale exhausted air and fresh intake air.

heat transfer fluid   The fluid circulating through a solar hot water collector, commonly called solar fluid.

high-grade energy   A relatively ordered form of energy, such as electricity.

high-head hydroelectric power plant   A hydroelectric plant utilizing a large vertical drop in water level to generate electricity, generally relying on a head of over 1000 feet.

high-quality wind   A strong, steady wind without turbulence.

hopper window   A type of window that opens into the living space by pivoting along the top or bottom surface.

horizontal-axis wind turbine   A wind turbine in which blades rotate about a horizontal axis, with blades moving perpendicular to the direction of the wind. These turbines must reposition the rotor to face into the wind.

hot, arid climate design   A building design characteristic of desert climates in which massive walls moderate daily temperature swings and ample shading prevents excessive heat gain during the day.

hot, humid climate design   A building design common to tropical regions where daily and seasonal temperatures remain fairly constant. Open and airy, with special attention to ample ventilation, these designs often include separate spaces for cooking and living.

HVAC   The heating, ventilating, and air conditioning systems included in a building.

hydraulic fracturing   The injection of water, sand, and corrosive chemicals at high pressure into deep crustal rock formations to create new fractures and widen existing fractures in order to improve the permeability of the formation.

hydroelectric   A form of renewable energy that converts the kinetic and potential energies of water into electricity.

hydro energy   A form of solar energy whereby evaporation and precipitation of water result in available gravitational potential energy.

hydrogenated chlorofluorocarbons (HCFCs)   Gases that include carbon, chlorine, fluorine, and hydrogen. Identified as being less damaging to the earth's ozone layer than chlorofluorocarbons (CFCs), they are commonly used as blowing agents for closed-cell foam insulation.

I

impact turbine   A turbine that rotates due to the impact of high-pressure jets of water focused on the rim of a wheel. Impact turbines are not submerged.

indirect gain system   A passive solar design that heats a surface exterior to the living space and uses natural forms of heat transfer to transport heat into the interior spaces.

insulated concrete forms (ICFs)   Interlocking foam blocks assembled at the building site and filled in place with concrete. The concrete core provides structural support, while the surrounding foam offers permanent insulation.

integral collector storage (ICS) system   A passive, open-loop solar hot water system in which the collector also serves as the hot water storage.

intertidal zone   An ecologically diverse and richly inhabited coastal region that is alternatively flooded and dry during natural tidal cycles.

inverter   A device that converts DC electricity into AC electricity.

isolated system   A sunspace or other type of passive solar design that is able to function independently of the building interior.

K

Kaplan turbine   A reaction turbine resembling a vertical propeller that rotates due to the weight of downward-flowing water across the blades.

kinetic energy   The energy of motion.

L

laminate flooring   A flooring material composed of layers of paper, wood, and resin bonded together under heat and pressure.

LED   Light-emitting diode bulbs are currently among the most efficient, durable, and long-lasting light sources available.

lift   A force caused by a difference in pressure on either side of a surface due to a difference in wind speed flowing across the surface. The force of lift acts perpendicular to the direction of wind flow.

lift-based design   A method of capturing wind energy using aerodynamic lift.

linoleum flooring   A flooring material composed of linseed oil, cork, wood flour, and powdered limestone bonded together under pressure.

load analysis   A method of determining electricity consumption by adding up the electrical usages of all items connected to the system.

long-lived   One of the six green building principles that compares the embodied energy and resource consumption of a structure to its longevity.

loose-fill insulation   Insulation in the form of granules, fibers, or other small pieces that can be poured, sprayed, or pumped into cavities. These include vermiculite, cellulose, and fiberglass.

low-E (low-emissivity) coating   A coating that reduces the amount of infrared radiation (heat) that passes through the glass panes of a window.

low-grade energy   A relatively disordered form of energy, such as thermal energy.

low-head hydroelectric power plant   A hydroelectric plant utilizing a relatively small vertical drop in water level to generate electricity, generally relying on a head of less than 100 feet.

M

manifold   A pipe that connects individual evacuated tubes and transfers heat from the tubes to the solar fluid passing through the manifold.

medium-density fiberboard (MDF)   A type of fiberboard made of wood particles and fibers mixed with adhesives and pressed into sheets.

medium-head hydroelectric power plant   A hydroelectric plant utilizing a moderate vertical drop in water level to generate electricity, generally relying on a head of between 100 feet and 1000 feet.

metal roofing   A type of roofing composed of coated steal, aluminum, or copper, that can take the form of shingles or long sheets.

metal siding   A type of siding, typically manufactured from steel or aluminum, that is light and durable, and can be recycled at the end of its useful life.

micro hydro   A small-scale hydroelectric facility with a generating capacity that varies from less than 100 watts to more than 100 kW.

mineral wool insulation   Insulation composed of blast furnace slag or minerals like basalt that is spun into fine fibers. It is available as rolls, batts, or loose fill.

monocrystalline solar cell   A solar cell created from a single crystal of silicon selectively doped and sliced into extremely thin wafers.

multiple panes   Windows composed of two or more parallel panes of glass, each separated by a narrow space filled with low-conductivity gas, such as argon.

municipal solid waste   City-collected garbage that consists primarily of paper, yard waste, food scraps, and wood products but also includes materials derived from fossil fuels such as plastic and rubber.

N

n-type (negative) semiconductor   A layer of silicon doped with phosphorous, creating extra electrons.

neap tide   The minimum tide that occurs each month during a first or last quarter moon when the sun and moon act on the earth from perpendicular directions.

net metering   A system in which a utility company credits the customer with the full retail value of the electricity generated by the customer's renewable energy system.

nuclear energy   The energy associated with the number of protons and neutrons in the nucleus of an atom.

O

ocean current   The continuous large-scale flow of water within the ocean along a steady path. Near-surface ocean currents are initially set in motion by the wind and influenced by the rotation and topography of the earth.

off-grid   A system that is independent of the electric utility grid.

Ohm's Law   A rule governing the relationship between the electrical current, voltage, and resistance: Current = Voltage/Resistance.

open-loop direct system   An active, open-loop solar hot water system that utilizes a small, variable-speed electric pump to move domestic water through a collector and into a solar storage tank.

open-loop solar hot water system   A type of solar hot water system that directly transports domestic water through the collector.

operational wind speed   The range of wind speeds between the cut-in and cut-out speeds over which a wind turbine generates power.

oriented strand board (OSB)   A structural board produced by chipping logs into strands, cross-aligning the strands, and gluing the layers together at high pressure.

overshot wheel   A type of waterwheel in which water falling from above strikes buckets and rotates the wheel using both the impulse of the water and its weight.

P

p-n junction   The boundary between n-type and p-type semiconductors.

p-type (positive) semiconductor   A layer of silicon doped with boron, creating a deficit of electrons.

parabolic trough collector   A solar collector consisting of trough-shaped parabolic mirrors that track the sun's east-to-west movement and concentrate sunlight onto fluid-filled absorber pipes.

parabolic trough power plant   A power plant that collects solar energy in long lines of parabolic mirrors that heat water to create steam and drive electrical generators.

parallel   A wiring arrangement in which the total current is the sum of the currents of the individual modules, while the voltage remains the same as that of a single module.

particleboard   A structural board made of small wood flakes or particles glued together at high pressure.

passive annual heat storage   A passive solar design strategy applied to underground structures whereby solar energy warms the structure all year and enormous thermal mass prevents summer overheating and winter overcooling.

passive solar design   A design that makes use of seasonal changes in the path of the sun to heat (and sometimes cool) structures by natural, nonmechanical means.

passive solar hot water system   A type of solar hot water system that relies solely on natural thermosiphons to transport fluid, requiring no electrical or mechanical device.

passive stall   A method of limiting rotor speed by designing blades to create turbulence naturally at excessive wind speeds, inducing a stalled situation.

Pelton wheel   A common type of impact turbine used for high-head conditions.

penstock   A channel or pipe that delivers water from the water supply intake to the turbine, often including a control gate for regulating flow rates.

performance and durability   The degree to which a material satisfies its intended purpose and withstands wear and tear and decay.

phantom load   The electricity consumed by an electronic device when the device is turned off.

photosynthesis   A process in which green plants absorb carbon dioxide from the atmosphere, combine it with water, and use the energy of sunlight to create carbohydrates and oxygen.

photovoltaic array   A group of solar modules wired together.

photovoltaic device   A device in which a built-in asymmetry draws excited electrons away from their point of excitation and delivers them to an external circuit before they relax back into their ground states.

photovoltaic module   A group of individual photovoltaic cells assembled together and protected behind tempered glass.

photovoltaics (PV)   A form of renewable energy that directly converts sunlight into electricity.

picture window   A nonopening window that provides light but not ventilation.

pitch control   A method of limiting excessive rotor speed by rotating turbine blades to become more parallel to the wind direction than optimal, thereby reducing lift.

plastic fiber insulation   Insulation made from polyethylene terephthalate (PET) plastic (such as recycled plastic milk bottles) and formed into batts similar to fiberglass.

plug flow   A type of anaerobic digester consisting of a heated tank with a flexible cover that allows for biogas collection.

plywood   A structural board made of wood veneers glued together with the grain of adjoining layers at right angles to each other.

pollution   The discharge of harmful substances that lead to the contamination of soil, water, or the atmosphere.

polycrystalline solar cell   A solar cell made from molten silicon that is poured into a mold, cooled, selectively doped, and sliced into thin wafers.

polyisocyanurate insulation   Closed-cell rigid foam board insulation that uses a low-conductivity gas (typically a hydrogenated chlorofluorocarbon) to fill the cells and is usually backed with reflective foil.

polyurethane insulation   Closed-cell foam insulation that can be applied as an expanding foam or installed in rigid sheets.

Portland cement   Pulverized limestone and other materials baked at high temperature. Portland cement, which serves as a crucial ingredient in concrete and other building products, has a large embodied energy.

potable water   Water that is suitable for drinking.

powerhouse   The protective housing containing the turbine, generator, and regulatory controls for hydroelectric power generation.

power   The rate of using energy, commonly measured in watts.

prefinished materials   Materials finished or partially finished at the factory, including prepainted and engineered products.

pumped storage   A method of increasing the capacity of hydroelectric plants by pumping water that has already flowed through the hydroelectric turbines back into the reservoir.

PV power rating   The rated power output of a photovoltaic array under standard test conditions measured in kilowatts (kW).

pyrolysis   A thermal conversion process that heats combustible biomass and condenses the volatile components into a liquid fuel, called bio-oil.

Q

quality of installation   A factor that influences the amount of air infiltration associated with a window. High-quality installation minimizes air infiltration between the rough framing of the building and the window unit as a whole.

R

R-value   A measure of the resistance to heat transfer by conduction. Larger R-values provide greater insulation value.

radiation   The transfer of heat by electromagnetic waves, such as light.

rammed-earth tire   A construction method that utilizes discarded automobile tires packed tightly with soil and stacked in rows around a partially subterranean passive solar home.

rated wind speed   The wind speed at which the wind turbine begins to produce its rated power output. This is typically in the range of 25–35 miles per hour for commercial turbines.

reaction turbine   A turbine entirely submerged in water that employs curved blades to deflect the flow of water. In being deflected, water exerts a reaction force on the turbine, causing it to rotate.

renewable energy   A form of energy that is not depleted by its use, being replenished through natural processes at a rate at least as great as its consumption.

resistance   A material's opposition to the flow of electrons.

resource depletion   A measure of the sustainability and renewability of a building material.

resource efficient   One of the six green building principles that includes the selection and quantity of building materials, as well as water consumption and maintenance requirements.

rigid foam insulation   A type of building insulation, commonly available in sheets measuring 4 feet by 8 feet, that can be made from polystyrene, polyurethane, or polyisocyanurate foams.

roof mount system   A mounting system that anchors solar collectors to the roofs of buildings using rigid brackets that either tilt the collectors at some optimal angle or hold the collectors parallel to and a few inches above the roof.

rotor area (A)   The circle swept out by the rotation of a rotor: $A = \pi r^2$, where r is the length of each rotor blade.

rotor   The part of a wind turbine that extracts kinetic energy from moving air and converts it to the rotational kinetic energy of a turning shaft.

S

Second Law of Thermodynamics   In any conversion, the energy of a system always changes from a more ordered form to a less ordered form.

selective coatings   Any of various transparent films that can be added to glass to selectively block particular wavelengths.

semiconductor   A nonmetallic material, such as silicon, that allows electrons to move freely only when in an excited state.

series   A wiring arrangement in which the total voltage is the sum of the voltages of the individual modules, while the current remains the same as that of a single module.

sick building syndrome   An illness affecting building occupants caused by indoor pollutants, microorganisms, or inadequate ventilation.

silicon photovoltaic cell   A solar cell consisting of a thin layer of n-type silicon semiconductor joined to a thicker substrate of p-type silicon semiconductor.

site preparation   The process of removing trees and other vegetation, leveling the ground, and digging footings at a building site prior to construction.

slate tile   A smooth, flat sheet of slate made by splitting quarried stone along natural cleavage planes.

sod roof (living roof)   A roof consisting of soil or other growing medium established over a waterproof membrane that supports grass or other living plants.

solar chimney   A type of thermosiphon in which interior air is heated and escapes to the exterior, drawing in fresh air that provides ventilation and may cool a building.

solar fluid   The fluid circulating through a solar hot water collector.

solar heat gain coefficient (SHGC)   A value between 0 and 1 that measures the fraction of the sun's heat to pass through a window.

solar hot water   A form of renewable energy that converts the electromagnetic energy of sunlight into the thermal energy of hot water.

solar pathfinder   An instrument used to determine the degree of shading at a particular site.

solar rating and certification corporation (SRCC)   An organization that measures and certifies the performance of solar collectors and provides useful information on expected heat output.

solar resource   The available solar energy at a particular site.

solar thermal electricity   Electricity generated through the use of mirrors that focus the sun's energy to boil water and drive a steam turbine.

solar tower power plant   A power plant consisting of pole-mounted, dual-axis tracking mirrors that direct sunlight onto a receiving chamber in a central tower to boil water and drive a steam turbine.

solid wood flooring   A traditional type of flooring composed of solid wood that is natural, nontoxic, and can be recycled at the end of its useful life.

solstices   The dates when the sun's rays are the farthest from the equator, on or about June 21 and December 21, producing the longest day of the year in one hemisphere and the shortest day of the year in the other.

spray-in-place foam insulation   An aerated liquid that rapidly expands and solidifies to seal cavities, providing both conductive and convective insulation.

spring tide   The maximum tide that occurs each month during a full or new moon when the sun and moon act on the earth in the same direction.

stall control   A method of limiting excessive rotor speed by rotating turbine blades to become more perpendicular to the wind direction than optimal, causing turbulence to form behind the blades and reducing lift.

stalling   A process in which lift is destroyed by causing turbulence behind the blades.

standard test conditions   A solar intensity of 1000 watts per square meter at a temperature of 25°C (77°F). This condition approximates full sun at sea level with the module facing directly toward the sun.

stone flooring   A type of flooring consisting of natural quarried stone. Stone flooring is natural, durable, fireproof, and nontoxic.

stover   The nongrain portion of a corn plant, including the stalks, leaves, and husks.

straw-bale insulation   A natural insulating material consisting of untreated bales of straw. Lacking any fire retardant or mold inhibitor, straw bales must be kept dry and away from sources of ignition during construction.

structural insulated panels (SIPs)   A construction method using panels of rigid foam sandwiched between two sheets of oriented strand board. The panels provide both structural support and insulation for the building, replacing studs and cavity insulation.

stucco/plaster   A type of siding made of Portland cement, lime, and sand that is usually applied and allowed to dry in place.

sunspaces   A passive solar system, such as solariums, conservatories, and attached greenhouses, in which south-facing spaces share heat with an adjoining building.

synchronous inverter   A device that converts DC electricity into AC electricity, while matching the voltage and frequency to the utility grid.

syngas   A synthesis gas, formed by gasification, that consists primarily of hydrogen and carbon monoxide.

T

tailrace   A means for water to leave quickly and easily without backing up or interfering with the turbine after providing power for a hydro facility.

temperate climate design   A building design suitable for most regions in the United States that minimizes heat loss in the winter and protects from excess heat gain in the summer.

thermal break   Any nonconductive material that separates the inside of a window frame from the outside and reduces conductive heat transfer.

thermal conversion   A process of heating combustible biomass with too little oxygen to allow burning. Thermal conversion processes include charcoal manufacturing, gasification, and fast pyrolysis.

thermal energy   The kinetic energy of randomly moving atoms and molecules within a material.

thermal mass   A massive material that can store large amounts of thermal energy.

thermal storage wall   An indirect gain passive solar system in which a wall collects and stores solar energy during the day and releases it gradually to the interior during the night.

thermosiphon   A natural passive solar system that utilizes the thermosiphon effect. As an indirect gain passive solar system (Chapter 3), a thermosiphon utilizes solar heating to drive convective motions that can provide heating, ventilation, and/or cooling to a building. As a type of solar hot water system (Chapter 6), a thermosiphon consists of an insulated tank mounted directly above a collector.

thermosiphon effect   The natural tendency of hot fluid to rise and cold fluid to sink due to differences in density.

tidal barrage   A shallow dam constructed across an inlet or estuary.

tidal energy   A nonsolar form of energy resulting from the gravitational interaction of the moon and sun on the earth's oceans.

tidal stream   An ocean current resulting from a tidal cycle.

timber frame   A construction method that replaces standard dimensional lumber with large timbers custom cut to fit together with wooden dowels or pegs. Also called post-and-beam construction.

tinted glass   An exterior coating applied to glass that reflects light, reducing transmittance. This is usually done to reduce solar heat gain or increase privacy.

toxicity   The degree to which a material is poisonous or damaging to an organism.

tracking array   A mounting system in which the array tracks the sun's path to maximize power output.

transesterification   A process by which alcohol chemically reacts with vegetable oil to produce biodiesel and glycerin.

trombe wall   A thermal storage wall.

turbulence   Rapidly fluctuating flow that varies chaotically in speed and direction.

type of seal   Elastic seals between the sash and the window frame restrict the air infiltration that allows convective heat transfer. Compression seals on hinged windows admit less air than seals around sliding sashes.

U

U-factor   A measure of the conductive insulation value of a window. Lower U-factors provide better insulation value: U-factor = 1/R-value.

undershot wheel   A type of waterwheel in which numerous blades dip into a flowing stream and the water pressure against the blades rotates the wheel.

uniform flow   A steady wind that is free of turbulence and sudden changes in direction.

V

vermiculite insulation   Small, lightweight pellets made by rapidly heating silicate minerals. Once commonly used as loose-fill insulation, concerns over asbestos limit current use.

vertical-axis wind turbine   A wind turbine in which blades rotate about a vertical axis, with some blades moving toward the wind, while others move away. These turbines harness wind from any direction without repositioning the rotor.

vinyl flooring   Made from polyvinyl chloride (PVC), vinyl is a common type of flooring due to its relatively low cost and minimal maintenance requirements. However, its production and disposal continue to pose environmental concerns.

vinyl siding   A relatively inexpensive and low-maintenance siding, vinyl has been popular for conventional residential construction. However, its production and disposal continue to pose environmental concerns.

visible transmittance (VT)   The amount of visible light to pass through a window, given as a percentage between 0% and 100%.

volatile organic compounds (VOCs)   Organic chemicals that readily evaporate at room temperature. Many, such as formaldehyde, pose risks to human health.

voltage   The pressure compelling an electron to move.

W

water supply   The source of water for a hydroelectric system. This can vary from a seasonal creek to a large reservoir or lake.

wave energy   A form of solar energy created through the action of wind on large bodies of water, resulting in available kinetic and potential energy.

wind energy   A form of solar energy whereby large-scale variations in planetary heating result in the available kinetic energy of moving air that can be converted into electricity or mechanical work.

wind quality   A measure of the suitability of wind conditions for energy generation.

wind speed (*v*)   The rate at which air moves past a point.

wood siding   A type of siding made of solid wood that has been used in various forms for hundreds of years.

wool insulation   A type of insulation made from natural sheep wool that has been treated with boric acid. Wool can be manufactured into either loose-fill or batt insulation.

Index